4TH
ORIGIN

Other books by Michael Ebifegha:

*Farewell to Darwinian Evolution: Exposition of
God's Creation Patent & Seal*

The Darwinian Delusion: The Scientific Myth of Evolutionism

*Creation or Evolution? Origin of Species in Light of
Science's Limitations and Historical Records*

*Satan's Shadow in Abrhamic Religions:
Clerics' defiance of God's Creation Sabbath Day mandate in
celebrating Charles Darwin's Evolution Day in their
places of worship*

4TH ORIGIN

Refuting the Myth of
Evolutionism
and
Exposing the Folly of
Clergy Letters

MICHAEL EBIFEGHA

Leavitt Peak Press

Printed in the United States of America.

Library of Congress Control Number: 2022946727

ISBN: 978-1-957956-91-6 (sc)
ISBN: 978-1-957956-92-3 (e)

Rev. date: 03/13/2023

To

The Lord

Jesus Christ

Reminding the Church in Revelation 3:1–3; 14–16 (NIV):

These are the words of him who holds the seven spirits of God and the seven stars. I know your deeds; you have a reputation of being alive, but you are dead. Wake up! Strengthen what remains and is about to die, for I have not found your deeds complete in the sight of my God. Remember, therefore, what you have received and heard; obey it, and repent. But if you do not wake up, I will come like a thief, and you will not know at what time I will come to you.

These are the words of the Amen, the faithful and true witness, the ruler of God's creation. I know your deeds, that you are neither cold nor hot. I wish you were either one or the other! So, because you are lukewarm–*neither hot nor cold*–I am about to spit you out of my mouth.

Contents

Preface

We urge school board members to preserve the integrity of the science curriculum by affirming the teaching of the theory of evolution as a core component of human knowledge. We ask that science remain science and that religion remain religion, two very different, but complementary, forms of truth.[1]

—The Evolution Clergy Letter

We urge school board members to fight for the integrity of our science curriculum by insisting that evolution's mortal flaws be published. We ask that science remain science so that truth may remain truth.[2]

—The Creation Clergy Letter

Indoctrination, either through religion or science, encroaches on liberty! It is not the role of public schools or the state to indoctrinate students with either *a specific religious belief (creationism) or a specific scientific belief (evolutionism). Let science remain within its empirical limits and be free of philosophical preferences.*

—Michael Ebifegha

The world is plunged into an endless controversy because modern scientists are engaged in a philosophical dispute about evolution as evidence of *mechanistic non-intelligence* and *chance* versus creation as evidence of *divine intelligence* and *design*. They redefine science in an effort to distinguish nature from God without prior knowledge of the origin of either. Dissent within the scientific establishment culminated in court battles. In the face of stubbornness on the part of comparatively few creationists in the scientific community, the evolutionist camp in 2006, under the influence of atheist Michael Zimmerman, extended the battlefield to churches for partisan support. That overture has generated controversy in the religious community, and the two warring camps, like their affiliates in the scientific establishment, are seeking endorsements from the public at large. Today, many mainstream churches in America observe either an Evolution Sunday close to Charles Darwin's birthday to profess their acceptance of his theory or a Creation Sunday to indicate their opposition to it.

This book is not intended to fuel the ongoing competition for signatures because God does not need a consensus. God knew that at some point in world history human beings made in divine image would address the topic of their species' origin, even musing on whether a piece of clay ever questions what the potter does (Isaiah 45:9). In making the claim to have created the earth and humankind (Isaiah 45:12) one of the commandments in the Decalogue, God accepted this fact as a moral issue. As a moral commandment, the mandate is to propagate the timeless truth and not disciplinary or philosophical preferences. The objective of this book, therefore, is to illuminate God's expressed truth, as opposed to refereeing the political debate between clerics and scientists regarding our species' origin. Nonetheless, the most productive way of delineating the truth is to respond to statements issued by the conflicting camps and relate them to God's personal claim in Scripture without favouring either church tradition or scientific preference.

Scientists focus on evolution because they perceive that creation presupposes a creator. To accommodate an evolutionary worldview, secularly minded clergy equate God's Creation Sabbath commandment to a Jewish custom and dismiss the Genesis story as an ancient myth. There is also the problem of a serious disconnection between clerics' understanding of evolutionary science and their interpretation of the Holy Scriptures. For instance, the "Letter from American Christian Clergy" contends that to reject the theory of evolution as foundational scientific truth is

- to embrace scientific ignorance and transmit such ignorance to our children;
- a rejection of the will of God;
- an attempt to limit God; and
- an act of hubris.[3]

This book explores the wealth of veridical evidence in the Holy Scriptures that challenges the validity of these specious assertions. It offers the following point-by-point rebuttals to the above views. To accept the theory of evolution as foundational scientific truth is

- to embrace ignorance of (1) God's foundational law on creation that defines our seven-day weekly cycle; and (2) God's designation as the embodiment of truth and the author of science;
- a rejection of God's will expressed in the Decalogue that stipulates six days of work (Sunday through Friday) to parallel God's six days of creation and a seventh day of rest (Saturday);
- an attempt to limit God's omnipotence and omniscience by proposing millions of years of evolution based on nothing, chance and ignorance instead of six days of creation based on divine power, wisdom and understanding (Jeremiah 10:12); and
- an act of hubris by insinuating that God in the Creation Sabbath law cannot distinguish between six days of creation and millions of years of creation or evolution.

The fact that the religious community is getting deeply involved in the evolutionism-creationism controversy is proof that Darwin's theory has more to do with religion than science. His revolutionary book titled *The Origin of Species* (1859) bears the wrong title. Distinguished evolutionist Ernst Mayr, professor of Zoology at Harvard University, reminded the world of this fact when he wrote, "It is quite true, as several recent authors have indicated, that Darwin's book was misnamed, because it is a book on evolutionary changes in general and the factors that control them (selectivity, and so forth), but not a treatise on the origin of species."[4] While diversity of species is a scientific component related to breeding, the origin of species is a religious component that parallels special creation in Genesis. The religious component makes Darwin's theory unique from other scientific theories. Philosopher Michael Ruse, an ex-Christian, affirms that "Evolution (evolutionism to be exact) is religion. This was true in the beginning, and it is true of evolution(ism) still today....Evolution(ism)... came into being as a kind of secular ideology, an explicit substitute for Christianity, with meaning and morality."[5]

An historic step toward achieving the goal of making evolutionism an explicit substitute for creationism occurred on February 12, 2006, which marked the 197th anniversary of Darwin's birth. That Sunday was observed in some churches as Evolution Sunday. On such occasions Darwin's theory of evolution is explored in lessons, sermons, and discussion groups. A majority of Christians ignore God's Saturday memorial of creation, instead setting aside Sunday to commemorate Christ's post-resurrection appearances and the descent of the Holy Ghost on Pentecost. But on Evolution Sunday evolutionism triumphs over creationism and marginalizes Christ's resurrection appearances and Pentecost. God's six days of creation and seventh day of rest, the cornerstone of the world's weekly cycle, are replaced by Darwin's millions of years of evolution under the supervision of natural selection and mindless mutation. Human beings were therefore not created directly in the image of God in accordance with the Scriptures, but instead evolved from bacteria, sharing an unknown common ancestor with chimpanzees. This is what some modern Christian clerics present as foundational scientific truth to their congregations. The recognition of Darwin's impact on religion is the reason why there is a Darwin Day celebrated by the secular scientific establishment and now by the liberal religious community. If solid and universally indispensable contributions to science were the criteria for such celebrations, Isaac Newton and Albert Einstein would have been better choices.

Because no empirical evidence validates molecule-to-human transitions, clerics cannot authoritatively assert that the theory of evolution is a foundational scientific truth. The foundation of Darwin's theory of evolution, unlike cell theory, is based solely on authority; the theory has no rigorous empirical background. Under cell theory, cells are unified, whereas under evolutionary theory species are diversified, with countless branches diverging from a common ancestor, such that truth is foreign to the paradigm. Empirical investigations presented at the 2016 Royal Society's "New Trends in Biological Evolution" conference are challenging the Neo-Darwinian paradigm by showing that new species arise through *cooperation and rapid change* as opposed to *competition and gradual change*.[6] In other words, Neo-Darwinism does not produce new species, but Anti-Darwinism does. Whereas this may not be good news to evolutionists, it is comforting to creationists who advocate the literal interpretation of Genesis cosmology, which reveals the rapid production of various life forms without competition among the creatures. Evolutionist James MacAllister, curator of the Lynn Margulis archive at the University of Massachusetts Amherst, who wrote a thorough review of the Royal Society evolution conference conveyed his frustration with the ground-breaking title, "The Royal Society's Evolution Meeting: James MacAllister/Why Neo-Darwinism Was the Biggest Mistake in the History of Science."[7] We still have much more to ponder.

Recent issues of journals such as *New Scientist* devoted to the topic "Darwin Was Wrong[:] Cutting Down the Tree of Life" (24-30 January 2009), and books by accomplished scientists and philosophers such as Jerry Fodor and Massimo Piattelli-Palmarini's *What Darwin Got Wrong*

(2010) confirm that Darwin's theory is not a foundational scientific truth. *New Scientist* has no religious axe to grind, and both Fodor and Piattelli-Palmarini are diehard atheists who believe in evolution. We thus can infer that Darwin Day is a partisan effort designed by his supporters to silence genuine scientific opposition to the ailing theory of evolution by means of natural selection.

The modern scientific establishment may be able to subdue the efforts of individual authors or journals who dare to suggest that Darwin was wrong, but it is unable to silence resilient creationists or intelligent-design scientists. This fact prompted evolutionist Jerry A. Coyne, author of *Why Evolution Is True* (2008), to remark that "During more than twenty-five years of teaching and defending evolutionary biology, I've learned that creationism is like the inflatable roly-poly clown I played with as a child: when you punch it, it briefly goes down, then pops back up."[8] No assault, however, can nullify the truth of our origin. Intelligent design affirms God's law and word as the truth (Psalm 119:142, 160).

God's truth is complete because it embraces both the material and nonmaterial realms. The limits of science in this regard are well understood. According to Eugenie C. Scott, executive director of the National Center for Science Education, "Science is nothing if not practical. The explanations that are retained are those that work best, and the explanations that work best are ones based on material causes. Nonmaterial causes are disallowed."[9] Human beings are characterized by both a material component such as the *brain* and a nonmaterial component such as the *mind*. To dismiss the *mind* as irrelevant is to question the credibility of science as an exclusive evidentiary path to truth on the subject of human origin. The *origin* of species, as opposed to the diversity of species, relates to a singular event; hence, it falls outside science's purview because the event cannot be tested or repeated. No amount of scientific explanation based on material causes can establish the truth about human origin. Physicist H. S. Lipson, Fellow of the Royal Society, while discrediting the creation account in Genesis as crude, contends that creation is the only credible answer to the question of origins. He argues that as a scientist, he is not happy with this conclusion, but he finds it distasteful for scientists to reject it altogether because it does not fit their preconceived ideas.[10]

Why bother, then, with a scientific explanation based exclusively on evolution? Physicist Richard Feynman's proposition, "I think it's much more interesting to live not knowing than to have answers which might be wrong"[11] is preferable for those who seek nothing but the truth. The problem is that the choice of wanting not to accept anything less than truth is currently denied because evolutionism has become a scientific religion, and many scientists are prepared to bend their observations to fit it.[12] It also is not the case that no theory of creation parallels the theory of evolution.[13,14] Today, all productive empirical scientific research lies within the species boundary and is preferentially classified as microevolution. Consider, for example, that without microcreation (events such as cell division or protein biosynthesis) there would be no microevolution (events such as mutation). Is microcreation less scientific as a natural process than microevolution? It is just that evolutionists do not want to embrace any theory of creation because it would point to a creator, a truth they do not want to accept. Accordingly, evolutionists detest intelligent-design theory and have convinced many liberal clergy to protest its presentation in public schools. It is every one's right either to accept or to reject evolutionism presented as a valid theory on the origin of species.

The church does not fall under the jurisdiction of the state, but public schools do. On the subject of human origin, evolutionism has as much indoctrinating influence or power as creationism. According to the U.S. Supreme Court's 1992 declaration, "At the heart of liberty is the right to define one's own concept of existence, of meaning, of the universe, and of the mystery of human life."[15] Human origin is an interdisciplinary topic involving history, religion, and science. Origin studies should therefore not be subject to disciplinary boundaries; and the choice of which worldview to embrace belongs essentially to the individual and must not be *deliberately* influenced in public education. Making choices in an academic setting involves exploring every possible explanation of given data and comparing the relevant interpretations. In the course of this process some may appear reasonable and others may not. Science, for example, cannot determine the structural age of the earth, just as it cannot determine the structural age of a building. So if some scholars contend that the earth is six thousand years old while others posit that it is 4.5 billion years old, the values have no meaning because there is no known framework of reference. No one observed the beginning of time. No educator, scientist, cleric, or judge therefore has authoritative knowledge on the subject. This fact means that school boards and courts of law cannot rule in favour of any particular worldview.

Today, evolutionism is imposed on students through the science curriculum, denying them the liberty to compare it with other worldviews. Little wonder, then, that the modern scientific establishment is flooded by scholars who have been taught only the evolutionist perspective. Banning the teaching of creationism or intelligent design is to embrace scientific, social, and judicial ignorance.

To limit or restrict God in any way for partisan gain is a consummate act of *hubris*.

This book is a guide for those who love the truth and perceive religion and science as God's way of communicating the same truth through different avenues.

Why This Book Matters

Scientific Theory Based on Authority and Just-So Stories

But what counts as fossil evidence for a major evolutionary transition? According to evolutionary theory, for every two species, however different, there was once a single species that was the ancestor of both. We could call this one species the "missing link." As we've seen, the chance of finding that single ancestral species in the fossil is almost zero. The fossil record is simply too spotty to expect that.[1]

—Jerry A. Coyne

Scientific Theory Based on Evidence

Science is timeless and is true when the events addressed are testable, repeatable and reproducible. A foundational or true scientific theory therefore cannot be developed based on essentially zero data or absence of evidence. As science of the gaps, missing link connotes missing science.

Therefore, if creation or intelligent design are brushed aside as unscientific, then evolution or unintelligent design must also be pulled aside as unscientific.

—Michael Ebifegha

The public is confused because two groups of accomplished scientists—evolutionists and creationists—are examining the same data and arriving at two radically different philosophical conclusions. Both intend to explain the trajectory of earth's development by

---

[1] Jerry A. Coyne, *Why Evolution Is True* (New York: Penguin Books, 2009), 34.

examining the origin and diversity of present and past life forms. Because of the ongoing conflict, evolutionists are rallying clerics to support their platform. In compliance with these evolutionary principles, thousands of clerics now engage in an "Evolution Weekend" celebration honouring the nineteenth-century naturalist Charles Darwin. This affirms the view that the Darwinian paradigm of evolution is sustained by authority.

The data generating endless controversy within the scientific establishment are, interestingly, unambiguous. When scientists examine the living world and the fossil record in sites such as the Burgess Shale in Canada or Chengjiang in China, they find abrupt gaps separating the primary life forms, with no intermediate levels or continuous transitional stages. Evolutionists interpret these gaps as evidence of missing links that justify Darwin's theory of evolution by natural selection—a theory that claims all life forms have descended from some unknown primordial ancestor. Creationists, in contrast, interpret these gaps as fixed boundaries imposed by a designer to limit diversity and forbid transformations between the primary life forms. Science is neutral because it cannot establish the origin of God or the origin of nature (the latter is mindless but obeys physical laws). So, the difference in opinion within the scientific establishment is philosophical and not scientific. The goal is to explain who we are, where we came from, and where we are going.

Which scientists are right? Science is bound to neither theistic nor atheistic principles; therefore, to identify truth instead of disciplinary preference, let us examine the facts in this book with an open mind.

Acknowledgment

My profound gratitude to Yahweh/Jehovah/Allah, our Creator and Heavenly Parent, who is patient and always welcoming regardless of our disposition.

I extend my gratitude to all who will find time to read part or this entire book and share their views whether positive or negative with others.

I am greatly indebted to all the authors cited in this book. Their works are the pillars of this book. The words of encouragement from Matthew Adeoye and Grace Nnamdi are sincerely acknowledged.

The editorial assistance of Dr. Robert Synder of www.edit911. com is greatly acknowledged; my thanks to Leavitt Peak Press for the marvelous job in preparing this book.

I sincerely thank my wife Margaret and daughters Mary-lyn, Kendella, Mercy, and Michelle for their support. Special thanks to the Holy Ghost Devotees Church founded by my late father Godfrey Allison Ebifegha, and the entire Amah, Ebifegha, and Fiebai families.

Introduction

If all the animals and man had been evolved in this ascendant manner [bacteria-to-human transformations], then there had been no first parents, no Eden, and no fall. And if there had been no fall, then the entire historical fabric of Christianity, the story of the first sin and the reason for an atonement, upon which the current teaching based Christian emotion and morality, collapsed like a house of cards.[1]

—H. G. Wells.

It is to challenge any such preposterous conclusion that through an unprecedented supernatural display of power and authority, the omnipotent and omniscient God intervened in human history to personally and publicly claim credit for having created the world; made the claim an everlasting commandment, a sign of holiness and a covenant of trust; and revealed God's Creation Sabbath Day as the timeless seal of ownership and divine immunity to any competing worldview.

A foundational scientific truth is timeless, so if the evolution worldview is the foundational scientific truth, then the absence of transitional stages or intermediate species that we observe in the living world today is false. Science and experiential knowledge, accordingly, are both myths. Whoops! Since this is not true, then, the evolution worldview must be false!

—Michael Ebifegha

At a time when atheists are ruthlessly and fraudulently using science to consolidate their conviction, the liberal clergy of our era in response are compromising the reason for their ordination and the foundation they inherited from their predecessors. Christianity is built on the platform of creation and on the foundation of the apostles and prophets with Christ Jesus as the cornerstone (Ephesians 2:19-20). Evolution, inaccurately defined as change with time, like every other secondary natural process such as growth or erosion, is a consequence of creation, and its goal is to modify things already in existence. It does not and cannot create new things from scratch. The Darwinian paradigm of evolution by natural selection that alleges numerous transformations resulting from a self-replicating molecule has not been proven and hence cannot constitute a foundational scientific truth, as some modern clerics are teaching their congregations.

Science relies on evidence that is empirically observable, testable, and repeatable. When certain events do not meet these criteria, such as bacteria-to-human transformations, then the explanations must include all possible options, whether or not they lead to theism, agnosticism, or atheism. The issue of whether extant evidence indicates intelligent or non-intelligent design is philosophical and not scientific.

Scientific laws are established facts that are timeless and theories are the correct explanations of these laws. Students are not compelled to *believe in gravity* because the *laws* and *theory* of gravity constitute timeless foundational scientific truths that can be derived, repeatedly tested and validated. Darwinian transformational evolution as a secondary process addresses a hierarchy of changes with time with no specific boundaries;

the alleged changes cannot be tested or reproduced. Darwin's theory of evolution deliberately undermines the scientific law of biogenesis. In comparison to the law of gravity, natural selection in biological evolution is a mindless process that is driven by environmental factors and not a unique event that can be quantified or formulated into a timeless scientific law. Because there are predominantly just-so explanations and stories that govern biological evolution, students are compelled in science classes to *believe in the Darwinian theory of evolution.* The scientific establishment is presently divided on the question of whether or not the Darwinian transformational evolution contradicts the laws of thermodynamics and blatantly defies information theory. Besides, the philosophical interpretations of key scientific evidence sometimes defy experiential knowledge. Consider this common-sense analogy. The complexity of animal and plant cells, with dimensions between 1 and 100 micrometers, is mind-boggling and marginalizes the abstract intelligence required to assemble from scratch the most sophisticated computer system. For evolutionists to attribute the existence or origin of the cell to a non-intelligent mechanism negates common-sense and experiential knowledge.

The unbiased scientist is neither an evolutionist nor a creationist, either of which is a religious designation, but will always humbly follow the evidence wherever it leads without imposing philosophical barriers. In other words, what makes *science* scientific is that it is neither theistic nor atheistic. Great scientists such as Albert Einstein and Richard Feynman never involved themselves in the evolutionism-creationism debate because, in order to protect science's integrity, they separated their philosophical convictions from their disciplinary practice. They respected the boundary between belief and science. In contrast, many scientists today have dissolved that boundary and convinced the media, politicians, and the judicial system that science is atheistic, such that theistic explanations of human origin must be excluded from science.

They insist that the natural world could only have been framed by chance and unintelligent mechanisms, yet all scientific achievements have involved designed plans as opposed to chance and carry the imprint of purposeful intelligence. To pick and choose what is or what is not science in order to promote a philosophical preference does nothing to enhance its integrity as an avenue of truth in understanding human origin.

Jonathan Wells, who holds a doctorate in molecular and cell biology from the University of California, Berkeley, and also a doctorate in religious studies from Yale University, writes:

> Evolution can mean many things. Broadly speaking, it means simply change over time, something no sane person doubts. In biblical interpretation it can mean that God created the world over a long period of time rather than in six 24-hour days. In biology it can mean minor changes within existing species, which we see happening before our eyes.[2]

Well continues:

> But Darwin's theory claims much more—namely, that all living things are descended from a common ancestor and that their present differences are due to unguided natural processes such as random variations and survival of the fittest. It is not evolution in general but Darwin's particular theory (Darwinism) that Evolution Sunday celebrates. That's why it is timed to coincide with Charles Darwin's birthday.
>
> The idea originated with University of Wisconsin evolutionary biologist Michael Zimmerman after a Wisconsin school board adopted the following policy in 2004: "Students are expected to analyze, review, and critique scientific explanations, including hypotheses and theories, as to their

strengths and weaknesses using scientific evidence and information. Students shall be able to explain the scientific strengths and weaknesses of evolutionary theory. This policy does not call for the teaching of Creationism or Intelligent Design."

Zimmerman called the policy a decision "to deliberately embrace scientific ignorance."... Not only did Zimmerman oppose analyzing Darwinism's strengths and weaknesses, but he also appealed to Christian churches for help. Why?

Polls have consistently shown that about 40 percent of Americans believe God created human beings in their present form a few thousand years ago, while another 45 percent believe that humans developed over millions of years from less advanced forms but that God guided the process. Despite their differences, both of these groups accept a central tenet of Christian theology: Human beings were designed and created in the image of God.[3]

The problem boils down to recognizing science's limits. Coyne asserts, "What moves science forward is ignorance, debate, and the testing of alternative theories with observations and experiments. A science without controversy is a science without progress."[4] This being the case, why are some modern scientists sensitive to students' objectively weighing the merits of Darwin's theory of evolution?

Scientists may defend their philosophical opinions, but real science does not require any form of defense.

The truth is that scientific evidence such as abrupt gaps in the fossil record and the living world can more readily be explained under the creation rather than evolution paradigm. Because science is evidence-based, scientists must explain and demonstrate empirically how life and the raw material that triggered the evolutionary process came into existence. Instead, however, people are urged to accept as foundational truth a theory that has many unanswered questions, and unlike traditional theories in science it comes with philosophical ramifications. Just as the public should not be bullied into accepting religious beliefs, so it should not be bullied into accepting pseudoscientific beliefs.

It is the objective of this book to point out to clergy and the public that science is not concerned with whether or not there is a Creator. Its duty is simply to sift the empirical evidence that points to truth. Scientific theories constitute explanations that can change with the emergence of new evidence. Clerics, then, should not presume to draw conclusions about matters that lie outside their area of disciplinary expertise.

Science and religion are never at war. It is not on scientific but rather on philosophical grounds that the natural law of biogenesis and DNA as the blueprint for life, collectively point to a Creator. Also, it is not on religious but rather on historical and civil grounds that the Creation Sabbath law in the Decalogue affirms God's designations as Creator and owner of the universe.

Genuine science, such as microcreation or microevolution, speaks for itself and warrants no debate within the scientific community. Macroevolution is a different story. Coyne defines the term as follows:

> "Major" evolutionary change, usually thought of as large changes in body form or the evolution of one type of plant or animal from another type.

> The changes from our primate ancestor to modern humans, or from early reptiles to birds, would be considered macroevolution.[5]

With this vague definition in mind, Coyne refers to the ruling of Judge John Jones III, a devoted churchgoer and conservative Republican, against the defense's claim in the Dover Area School District case that the theory of evolution was fatally flawed.

> To be sure, Darwin's theory of evolution is imperfect. However, the fact that a scientific theory cannot yet render an explanation on every point should not be used as a pretext to thrust an untestable alternative hypothesis grounded in religion into science classrooms to misrepresent a well-established scientific proposition.[6]

If testability is the definitive criterion for a scientific theory, no scientist has ever presented a test that demonstrates the transformation of a reptile to a bird. Darwin's theory of evolution would have made significant impact in the breeding enterprise if today we have; for instance, fruits that are partly apple and partly orange. Now, if fruits that have no minds (immaterial component) are limited to their boundaries, it is foolhardy to believe that organisms, which have both material and immaterial components, can transform from simple to complex life-forms. The supposed millions of years to produce major evolutionary changes in both the material and immaterial domains are outside the timeframe of scientific observation. The truth is that both macrocreation and macroevolution are outside science's limits and can only constitute myths. Both creationism and evolutionism are untestable beliefs.

Philosophical preference may prevail for a while; however, in the long run, only truth will set the record straight, and that is slowly happening. Evolutionists, for instance, now contend that "Transitions between species documented by the fossil record seemed to be abrupt, perhaps too abrupt to be explained by the modern synthesis."[7] Also, in September 2012 the "dead-genes" or "junk" DNA theory, which is one of the strongest reasons Coyne presents against the need for a Creator, was overturned.[8] These major setbacks are clear indications that key evolutionary postulates are not necessarily true, and hence, that the theory is fatally flawed. The late Stephen Jay Gould, who developed the theory of punctuated equilibrium to support Darwin's theory, affirmed this point when he wrote: "Evolutionary biologists in general are famous for their facility in devising plausible stories, but they often forget that plausible stories need not be true."[9] This point presumably explains why Zimmerman is uncomfortable with science students evaluating both the merits and the fallacies of evolutionary theory. Zimmerman's bias is reflected in the Clergy Letter Project where only the successes and not the numerous shortcomings of the theory of evolution are cited.

It may well be logical from the human point of view that mainstream clergy and theologians should align themselves with a majority opinion in the scientific community, even if this means compromising biblical truth, in order to avoid the Catholic Church's past mistakes in challenging scientific evidence that put the sun at the center of the solar system. The truth about God's creation of the world cannot be settled on the basis of popular opinion. The church was remiss in Galileo's case because it crossed the boundary between science and religion to challenge empirical evidence based on its misinterpretation of the scriptures. The church therefore was obliged to apologize belatedly in 1992.[10] In the evolutionism-creationism controversy, it is now the scientific establishment that is crossing over into religious territory on the basis of metaphysical assumptions. Because science cannot resolve questions about the origin of life or species,[11] its scholars are coming up with different philosophical interpretations of the same scientific evidence.

A resolution by clergy who support evolutionism reads, "We ask that science remain science and that religion remain religion, two very different but complementary forms of truth."[12] Unless such clerics are also scientists, then by virtue of their resolution they cannot endorse evolutionism as foundational scientific truth. In reference to Judge Jones's ruling in favour of evolutionism, Coyne's remarks that "scientific truth is decided by scientists, not by judges."[13] The same point applies to clergy.

Because such issues as the origin of life lie outside science's realm of investigation, some scientists are using achievements in other areas of human experience that do not relate to the evolutionary paradigm to promote their philosophical preference. No evolutionary scientist presents these achievements in science and technology, such as a computer, an airplane or light bulb, as evidence of creation. The motive of the hardened evolutionist is evident in statements by Pierre Teilhard de Chardin, who describes evolution as a "movement whose orbit infinitely transcends the natural sciences and has successively invaded and conquered the surrounding territory—chemistry, physics, sociology, and even mathematics and the history of religions.... [It] is a general condition to which all theories, all systems, all hypotheses must bow and which they must satisfy henceforward if they are to be thinkable and true. Evolution is a light illuminating all facts, a curve that all lines must follow."[14] To many of its advocates evolution is more a god than a subordinate natural process. Sadly, many clerics from different religious backgrounds are now literarily bowing to evolutionism.

The major opposition to evolutionism is intelligent design. The objective of the pro-evolution Clergy Letter Project is to weed out intelligent design from science classrooms. The paradoxical question looms: Why would scientists want to use abstract intelligence to explain what has been designed unintelligently by nature? Is this not tantamount to academic nonsense or delusion?

God as the Supreme Creator knows the difference between creation and evolution. The truth about our human origin is presented as a story in Genesis and classified as a law in Exodus. God affirmed the truth of Creation in speech before a live audience and in print on stone tablets. This truth is historically recorded and cannot be denied by pseudoscientific doctrines. Students should be allowed to weigh all the facts, historical as well as scientific, and reach a conclusion based on them. To do otherwise is to deprive them of their right to genuine knowledge. Freedom of inquiry as opposed to intimidation should be the rule. It is to restore such unbiased learning and to further scientific integrity that the present discourse is directed.

Each chapter begins with visionary quotations from various sources and/or my profound comments. Chapter 1 introduces both the pro-evolution and pro-creation documents from the Clergy Letter Project. Chapter 2 analyzes the evolutionism-creationism controversy in order to distinguish between preferences and truths in origin studies. Chapter 3 explains why Evolution(ism) is not a foundational scientific truth. Given this analysis, chapter 4 undertakes a diagnosis of the Clergy Letters to address their legitimacy in commenting on fields outside their authors' expertise. What are clergy telling their congregations on Evolution Sunday? Chapter 5 involves an analysis of a sermon presented on the 2014 Evolution Sunday. Chapter 6 then addresses the consequences of some clerics' acceptance of evolutionism as foundational scientific truth. In light of atheist Sunday assemblies that parallel Christians' Sunday worship services, it is evident that the controversy is not between religion and science but between theism and atheism. In this regard, the phrase "separation of church and state" needs to be revised. Chapter 7 then discusses a new substitute that encompasses all sorts of belief. The book concludes after providing in chapter 8 seven recommendations to clergy when dealing with crucial issues that relate to divinity.

1

The Evolution and Creation Clergy Letter Project

Clergy Celebrations

Today we celebrate Evolution Sunday with the theme this year of "A Different Way of Knowing." This is the 9th year that Christian but also Buddhist, Islamic, and Jewish faith leaders call attention to the nature of science and religion, and especially to an endorsement and indeed a celebration of evolution.[1]
—Jill Joseph, Evolution Weekend,
February 9, 2014

Divine Disapproval

See to it that no one takes you captive through hollow and deceptive philosophy, which depends on human tradition and the basic principles of this world rather than on Christ.
—Colossians 2:8 (NIV)

Evolution is a secondary event that could under certain circumstances modify things already in existence but not create new things from scratch; just like other natural events such as aging, evolution is the consequence of creation and certainly not the means by which the basic life forms were established. Clergy has nothing to celebrate over evolution, no need for either a pro evolution or pro creation Clergy Letters! Instead, in obedience to God's moral law on origins, clergy are enjoined to observe a memorial of God's six days of creation every seventh day (Saturday) of the week (Exodus 20:11-13, John 4:16; Acts 17:1-3).

To avoid events such as Evolution Sunday or Weekend, the mandate to observe God's Creation Sabbath Day came directly from God's lips and engraved on stone tablets by God's finger. Clergy must remain faithful servants of God and not messengers of modern scientists. The primary consequence of ignoring God's Creation Sabbath law in the modern world is that Evolution Sundays have fostered spiritual confusion for the average believer.

—Michael Ebifegha

The pro-evolution *Clergy Letter Project* is a project that embodies statements in support of the teaching of Charles Darwin's theory of evolution (*evolutionism*) and in opposition to the teaching of *creationism* in public schools and today collects signatures in support of letters from *American Christian, Jewish, Moslem, Unitarian Universalist, Buddhist* and *Humanist Clergy* (new addition)."[2,3] Clerics of different religious faiths have united in an unprecedented manner to remove the concept of God from public schools' science classrooms. The founder and director of the project, Michael Zimmerman, is an atheist.[4,5] There are six separate letters each titled "An Open Letter Concerning Religion and Science: A Christian Clergy Letter, a Rabbi Letter, a Unitarian Universalist Clergy Letter, a Buddhist Clergy Letter, an American Humanist Clergy Letter and a Clergy Letter from American Imams that is no longer listed on the official site.

The pro-evolution Clergy Letter contends that evolutionism and creationism are two very different but complementary forms of truth, which is true only if both are myths. The pro-creation Clergy Letter is a rebuttal.[6] The Christian version of the pro-evolution Clergy Letter, with 15,642 signatures as of May 15, 2020, has been translated from English into Spanish, French, and Portuguese. It reads:

> Within the community of Christian believers there are areas of dispute and disagreement, including the proper way to interpret Holy Scripture. While virtually all Christians take the Bible seriously and hold it to be authoritative in matters of faith and practice, the overwhelming majority do not read the Bible literally, as they would a science textbook. Many of the beloved stories found in the Bible – the Creation, Adam and Eve, Noah and the ark – convey timeless truths about God, human beings, and the proper relationship between Creator and creation expressed in the only form capable of transmitting these truths from generation to generation. Religious truth, in other words, is of a different order than scientific truth. Its purpose is not to convey scientific information but to transform hearts.
>
> We the undersigned, Christian clergy from many different traditions, believe that the timeless truths of the Bible and the discoveries of modern science

may comfortably coexist. We believe that the theory of evolution is a foundational scientific truth, one that has stood up to rigorous scrutiny and upon which much of human knowledge and achievement rests. To reject this truth or to treat it as "one theory among others" is to deliberately embrace scientific ignorance and transmit such ignorance to our children. We believe that among God's good gifts are human minds capable of critical thought and that the failure to fully employ this gift is a rejection of the will of our Creator. To argue that God's loving plan of salvation for humanity precludes the full employment of the God-given faculty of reason is to attempt to limit God, an act of hubris. We urge school board members to preserve the integrity of the science curriculum by affirming the teaching of the theory of evolution as a core component of human knowledge. We ask that science remain science and that religion remain religion, two very different, but complementary, forms of truth.[7]

Clerics have missed the salient point in their effort to embrace evolutionism as foundational scientific truth. The creation account in Genesis may be subject to different interpretations because it is presented as a story. In Exodus, God's personal testimony pertaining to creation and origin is presented to the world as a commandment law in the Decalogue; acceptance of the truth of creation is accordingly a moral obligation. Religious truths pertaining to creation and redemption that transform hearts can only have one proper or correct interpretation and, hence, should not constitute areas of dispute and disagreement within the community of believers. God's loving plan of salvation is for the descendants of Adam who was uniquely created in God's image. Clergies are ordained to proclaim, from generation to generation, the truth of God's six days of creation based on plan, limitless understanding, wisdom and power. Clergies under God do not have the due authority to endorse millions of years of evolution based on chaos, lucky chance and selection as foundational scientific truth. Also clerics can share but not impose any worldview, whether religious on scientific, on any one. The facts in favour and those against a scientific theory are equally important just as the synonyms and antonyms of a word are. Therefore, students should be the ones to establish whether or not creationism and evolutionism are two very different, but complementary, forms of truth.

Details of the pro-evolution Clergy Letters from other faith traditions are provided in the Appendix.[8-12]

In rebuttal the pro-creation Clergy Letter, with 345 Creation Letter signatures, and 85 Clergy for creation worldwide as of May 15, 2020, reads:

> Observable, testable, repeatable science has brought us many benefits and innovations. The founders of modern science were Creationists, "thinking God's thoughts after Him." Most of the disciplines within science were founded before Darwin or by scientists who actually rejected his theory. The Scientific Method itself is based on the idea that an orderly creation can be rationally understood because it was designed by an Intelligent Creator. Creationists today continue to practice normal, experimental science without need of evolution.
>
> Evolution is not observable, testable, repeatable science. It's a belief about the past, an atheist Just-So Story seeking to displace the divinely revealed Creation record. It's based on the flaw of naturalism, which begs that all problems must have a natural explanation, so God isn't needed. This stands directly at odds with the biblical claim that God's existence, eternal power, and Godhead are self-evident in

His Creation, for it excludes an Intelligent Creator from all consideration. Faulty assumptions lead to faulty conclusions! Sadly, statistics demonstrate that children taught Godless evolution as scientific truth reject religious truth wholesale! It's time to judge this tree by its fruit!

The Bible stands as the inerrant, revealed Word of God. As such, the conclusions and speculations of fallible, finite men should be weighed in light of the revelation of an infallible, infinite God—not the other way around. Let God be true and every man a liar!

Some claim to take the Bible seriously but actually hold man's word as their true authority, so long as it calls itself science; where it disputes the Genesis record, they denigrate the Word of God to mere Bible stories in the tradition of Aesop's fables. After swallowing the camel of the Resurrection and supernatural miracles, they strain at the gnat of an historical Creation week.

We do not follow cleverly devised fables. While the Bible is NOT a science textbook, the Word of God is true and accurate in all it records. Jesus affirmed the truth and authority of God's Word, mentioning Creation, Adam and Eve, Abel, Noah, and Jonah as matters of fact. Though some object that religious truth is of a different order from scientific truth, Jesus refuted this false dichotomy when He asked Nicodemus, "If I have told you earthly things, and ye believe not, how shall ye believe if I tell you of heavenly things?" (John 3:12). The very reason

Jesus literally died and rose again is a world cursed by the literal Fall of a literal Adam!

We the undersigned affirm the truth of a biblical, literal 6-day Creation and strongly discourage any Bible-believing Christian from endorsing or celebrating an Evolution Sunday. Evolution is a lie which undermines both biblical authority and the foundational basis of the Gospel. We urge churches to send a clear message of the enduring authority of God's Word by celebrating a Creation Sunday instead of the Clergy Letter Project's proposed Evolution Sunday. We urge school board members to fight for the integrity of our science curriculum by insisting that evolution's mortal flaws be published. We ask that science remain science, so that truth may remain truth.[13]

It is not the objective of this book to support either camp of the Clergy Letter Project because both have failed to make any reference to God's personal intervention to isolate the Creation Sabbath Day as the benchmark of six days of creation, the most profound and historically recorded event that testifies that the world did not evolve but was created by God. God's Creation Sabbath Day (Saturday) observance is not a Jewish custom but rather a creation ordinance that calls for celebration of divine ownership of the universe (Exodus 20:8-11, 31:16-17, Nehemiah 9:14, John 4:16). There will be no need for an Evolution or Creation Sunday celebrations if Christians follow God's verbal and written instruction concerning the origin of the world. The contemporary standoff between an Evolution Sunday versus a Creation Sunday only confirms a theologically fractured Christian foundation today. God's commandment law about creation is literal truth and the timeless bulwark against false worldviews such as evolutionism. By not making reference to the Creation Sabbath legislation and the granting of the Ten Commandments, both the pro-evolution Clergy Letter and the pro-creation Clergy Letter underestimate God's wisdom in personally and publicly addressing the subject of *Origin* and at the same time providing a timeless immunity against false worldviews through a moral and ceremonial commandment law. Without the Sabbath law in the Decalogue as the ultimate authority, the pro-evolution Clergy Letter versus the pro-creation Clergy Letter controversy parallels the evolutionism versus creationism controversy within the scientific establishment.

The creation paradigm has its own set of controversies because God's instruction to observe the Sabbath day as a creation ordinance is not followed. Based on the same Scriptures, there are four views on creation: Young Earth Creationism (Answers in Genesis), Old Earth Creationism (Reasons to Believe), Evolutionary Creation (BioLogos), and Intelligent Design (The Discovery Institute).[14] God is all knowing and declares the end from the beginning (Isaiah 46:10); God is not the author of confusion (1 Corinthians 14:33). It is to avoid these divisions that God personally issued the Creation Sabbath mandate in speech and in print, and classified it as a moral commandment (Exodus 20:8-13), a sign (Exodus 31:12-13, Ezekiel 20:12,20), and a covenant (Exodus 31:16, Isaiah 56:4-6). The next chapter will unfold key literal truths in the evolutionism-creationism controversy.

2

Seven Truths in Evolutionism-Creationism Controversy

God as the Embodiment of Truth

God is not a man, that He should lie.
>—Numbers 23:19 (NKJV)

Indeed, let God be true but every man a liar.
>—Romans 3:4 (NKJV)

Science Cannot Challenge God's Truth.

For as the heavens are higher than the earth, So are My ways higher than your ways, And My thoughts than your thoughts.
>—Isaiah 55:9 (NKJV)

I have made the earth, And created man on it. I— My hands—stretched out the heavens, And all their host I have commanded.
>—Isaiah 45:12 (NKJV)

Science as a Pathway to Truth

One of the fundamental truths in biological sciences is the law of biogenesis which stipulates that life comes from preexisting life, and this law is consistent with our experiential knowledge. This means life cannot evolve from non-life.

Science is timeless. Any natural law or phenomenon that is not operative at the present such as abiogenesis (life from non-life) was also not in existence in the past and will not be operating in the future. In science because the present is the key to the past and to the future, the absence of transitional fossils in the living world is the evidence that there were none in the past and there will be none in the future.

—Michael Ebifegha

The debate on the origin of species is not a choice between modern science and religion. The evolutionism-creationism controversy is a conflict between two camps of scientists based on different philosophical interpretations of the same scientific evidence. The only credible way to decide the conflict between evolutionists and creationists is to focus on truth and not on philosophical preferences. To achieve this goal, scholars must examine the evidence in light of science's laws and limitations, describe the data with the appropriate terminologies, avoid the use of unprovable assumptions but follow the evidence wherever it leads, and interpret the evidence in light of authentic historical records. These records include: fossil data, credible historical revelations and/or claims by agencies. To account for both the material and immaterial components in origin studies, seven truths are identified. These truths when taken into consideration will lead to the correct philosophical interpretation of the scientific evidence.

Truth #1: Science is limited.

Science can neither establish nor deny God's existence. Science may corroborate some of God's revelations in Genesis 1, such as "earth and water together from the start."[1] Presently, however, scientists are unable to explain the origin of life, and how the world could be fully functional just in six days; for similar limitations, earthworms are unable to comprehend how human beings are able to establish skyscrapers. The term *chance* in evolutionary biology implies lack of knowledge, and *million to billion in years* means that the event in question is outside the domain of science.

Science is limited in the investigation of species and their relationships. For instance, the brain is an organ that can be analyzed scientifically with our five senses. The mind is an invisible or immaterial, but conscious, substance or process that is outside the purview of our five senses and, hence, indefinable using pure science. Science therefore cannot be the sole determinant of truth in the investigation of biological systems. For example, when Albert Einstein passed away in 1955 his brain tissue was preserved and sent to different scientific laboratories for analysis. What was unavailable for analysis was his mind that developed the mass (m)–energy (E) equivalence $E = mc^2$, where c is the velocity of light.

Truth #2: Use of correct terminology.

Creation is a primary process (e.g., DNA replication or protein biosynthesis) and evolution a secondary process (e.g., changes due to errors in DNA replication). The battle line was drawn when scientists chose to create a field in microevolution and name it *evolution* but did not create one for microcreation, the raw material for microevolution, and name it *creation*. Just as microcreation cannot add up to generate macrocreation so also microevolution cannot add up to produce macroevolution. As natural processes, the debate is not between evolution and creation. Instead, it is between evolutionism (belief in macroevolution) and creationism (belief in macrocreation). When the focus is on origins, which are primary events, the choice of a worldview in science based exclusively on secondary processes such as evolution and random chance is bizarre.

The Scriptures declare that human beings were created directly in the image and likeness of God (Genesis 1:26–27). Hence, the human mind could not have evolved from other species. Clerics' references to evolution as a foundational truth need to be properly addressed. In their sermons, clergy must contend with the biblical account that it took only a split-second for God to transform the proud king Nebuchadnezzar's mind from that of a human being to that of a grass-eating mammal so that for seven years he lived among wild animals and ate grass like cattle before he was restored as king (Daniel 4:28–37). In the seven years that he lived as an animal, his hair grew like the feathers of an eagle and his nails like the claws of a bird, so he was afflicted in mind as well as body. Clerics must make a choice. Is this evidence of God's retrieving and restoring divine image in a person or is it evidence of macroevolution (change with time by natural selection) or instantaneous macrocreation by an illimitable God? Equally of concern is modern clerics' attempt to segregate scientific truth from religious truth. If science and religion originate from God, should they not both lead to a single truth and not two radically different truths? Under the guise of scientism, clergy cannot, on the one hand, contend that over millions of years life evolved by chance from non-life and, on the other hand, proclaim that with God everything is possible regardless of time.

The terms microevolution and macroevolution are also vague because there is no well-defined boundary between the two. The prefixes "micro" and "macro" signify magnitudes or levels of evolution and the length of time involved. In this sense, microevolution refers to small changes in a short time interval that should not be readily observable and macroevolution refers to large changes in longer time intervals that should be easily observable. In contradiction, however, evolutionists claim that what is observable is microevolution and that what is hardly observable is macroevolution, which is inconsistent with the normal scientific understanding of these terms. A more appropriate usage would be to replace the prefix "micro" with "intra" in describing evolution within the species boundary and to replace "macro" with "extra" for evolution outside the species boundary. At the intraevolution level, there are small changes (microevolution) as well as large changes (macroevolution) that result in different degrees of modification without transformation as observed in the breeding enterprise. On the other hand, the extraevolution level relates to transformations that result in the production of novelties that so far are beyond the reach of science. Under the proposed definition intraevolution and extraevolution are autonomous fields,[2] but microevolutionary processes are simultaneously macroevolutionary processes, a fact that is consistent with Ernst Mayr's preposition.[3] Intraevolution is a scientific fact, and all achievements of modern science are confined to this autonomous field of evolution, whereas extraevolution typifies the just-so events that have never been observed and hence cannot constitute a foundational truth in science.

Truth #3: The origin of life and species.

If "origin" means from scratch and not from the consequences of a previous event, then science essentially has no worldview concerning the origin of species. According to the scientific law of biogenesis, biological life cannot arise from nonliving matter but from a previously existing life. Initial biological life, as a conscious condition or state; can originate only from an embodiment that is characterized by a mind that is everlasting, superior and illimitable in comparison to biological mind, a view consistent with Albert Einstein's perception of God as "the illimitable superior spirit with superior reasoning power."[4] Hence, without correct knowledge of the origin of life that precedes the origin of species, it is impossible to formulate a correct scientific worldview concerning the origin of species.

Accordingly, the following resolution by Episcopalian clergy is completely misguided: "Be it *Resolved*, That the theory of evolution provides a fruitful and unifying scientific explanation for the emergence of life on earth, that many theological interpretations of origins can readily embrace an evolutionary outlook, and that an acceptance of evolution is entirely compatible with an authentic and living Christian faith; and be it further *Resolved,* That Episcopalians strongly encourage state legislatures and state and local boards of education to establish standards for science education based on the best available scientific knowledge as accepted by a consensus of the scientific community."[5]

Scientists are conscious of the fact that without prior knowledge of the origin of life, they cannot truthfully explain the origin of species, which is why over a decade ago, they quietly set up the Origin of Life Prize of $1,000,000 to award the scientist(s) who proposed a highly plausible *mechanism* for the spontaneous emergence of genetic instructions sufficient to give rise to life.[6] The Foundation's Website, last updated in November 2013, provides the following news:

> On October 26, 2013 the Governing Board of the Origin of Life Science Foundation, Inc. voted to put on hold the Origin of Life Prize Program and to temporarily suspend the Origin of Life Prize offer. Over the 13 years since the Origin of Life Prize was first announced *in Nature* and *Science*, no submission has ever made it past the screening judges to higher-level judges. No submission has ever addressed, let alone answered, any of the questions below, for which the Prize offer was instituted. Most of these Prize-offer questions centered on: "How did inanimate, prebiotic nature prescribe or program the first genome?"
>
> Life-origin literature continues to circumvent and ignore this problem, if not deliberately sweep it under the rug. The Prize Program did much to raise consciousness and stimulate more consideration of the real problem of life origin—prescription of

future biofunction that was not yet selectable by
the environment.[7]

The above is evidence that understanding the origin of life requires
superior intelligence, and so at the next level of research the origin
of species cannot simply be an interplay of random chance, natural
selection, and genetic mistakes.

In a discussion paper the Origin of Life Science Foundation
affirmed the following facts:

In all known phenomenological life, genetic code manifests

- the conveyance of a functional coded message, using a sign
  system, to distant sites through an information channel to
  energy-consuming decoding receivers—ribosomal "machines";
- symbolic, indirect representation of that message from one
  alphabet into another (e.g., codons of nitrogen-based "language"
  being translated into the end-product of physical amino-acid-
  sequence "language");
- pre-specification of extremely unlikely and complex future events
  suggesting "*apparent* intent," "*apparent* planning," or "*apparent*
  purpose" (as Richard Dawkins describes it, "apparent design");
- instructions capable of effecting and affecting many individual
  manufacturing processes, and of mediating the cooperation of
  all of those diverse processes toward the one organismal and
  seemingly "conceptual" end of being and staying alive;
- the ability of that information (instruction) not only to give
  the directions or orders of what should be done, but *to bring
  to pass* those orders in the form of actual physical molecules,
  products, and life processes;

- the seemingly "irreducible complexity" argued by Michael Behe and
- the initial writing of this prescriptive information by nature, not just the modification of existing genetic instruction through mutation.[8]

All of the above elements are components of intelligence in both planning and execution. To address correctly the issue of species' origin, these are just the prerequisites stipulated by modern scientists who distance themselves from intelligent-design peers by the following declaration:

> The Origin of Life Science Foundation should not be confused with "creation science" or "intelligent design" groups. It has no religious affiliations of any kind, nor are we connected in any way with any New Age, Gaia, or "Science and Spirit" groups. The Origin of Life Science Foundation, Inc. is a science and education foundation encouraging the pursuit of natural-process explanations and mechanisms within nature. The Foundation's main thrust is to encourage interdisciplinary, multi-institutional research projects by theoretical biophysicists and origin-of-life researchers specifically into the origin of genetic information/instructions/message/recipe in living organisms. By what mechanism did *initial* genetic code arise in nature?[9]

A foundational scientific theory must come equipped with its explanations and not violations of established natural laws and must be supported by a clear understanding of mechanisms involved; it appears the Darwinian theory of evolution is the only exception presumably because of its philosophical implications. Creationists have been complaining that the just-so stories are inadequate to sustain the theory of evolution as pure science. In this regard, Nobel laureate Ernst Boris Chain wrote:

> These classical evolutionary theories are a gross oversimplification of an immensely complex and intricate mass of facts, and it amazes me that they were swallowed so uncritically and readily, and for such a long time, by so many scientists without a murmur of protest.[10]

In order to discredit creationism, the Foundation publicly presents evolutionism as a scientific fact based on circumstantial evidence, but behind the scenes, it is engaged in serious research to address the truth that creationists accuse them of concealing or neglecting. Scientists usually are vocal about important projects, but the Origin of Life Science Foundation is determined to keep a low profile and allow no media interviews until after its Origin of Life Prize is won. Need I say more?

Truth #4: The true age of the earth is unknown.

Scientific methods of age determination are unreliable. By radiometric dating alone, science cannot distinguish between houses that are ten years old from those that are one hundred years old. Only correct eyewitness documentation on the property can provide the true age. For example, Mount St. Helens erupted on May 18, 1980, and created a new structure that hardened into rock in less than a decade. Radiometric dating of some constituents of the new rock formation gave a significantly different age from its correct structural age.[11] For the same rock sample, different laboratories would give significantly divergent estimations. This could be the result of several factors, such as the unproven assumptions made in the age analysis. As further evidence, scientific attempts to determine the chronological ages of people based on body components have been proven wrong upon the presentation of actual birth records.[12] God's six days of creation does not indicate how old the earth is. The structural age of a building is independent of the ages of its occupants. Therefore, attempts to deduce the earth's age based upon people's genealogical ages in the Scriptures is nonsensical. In their book titled *The Grand Design* (2010), Stephen Hawking and Leonard Mlodinow posit that humans are a recent creation because if people had been around for millions of years, they would be much further along in technological mastery.[13] Millions of years are only necessary for evolutionism to make any sense.

Truth #5: God's Creation Sabbath law.

After six days of creating the world, God reportedly added a seventh day as evidence of the supernatural completion of Creation and seal of ownership of the universe. The seven-day week has no definite astronomical significance unlike the year and months that respectively gauge the movement of the earth around the sun and moon around the earth. However, as a cosmological benchmark, the seventh day of rest isolates the beginning of the first day of creation from the end of the last day of creation. Without the seventh day block of holy time, the first day of creation will intersect the last day of creation; hence, a six-day weekly cycle will cosmologically be meaningless.

Paralleling the six days of creation and seventh day of rest, God then led the children of Israel after their escape from slavery in Egypt through the desert and, for forty years, supplied them with food dropped from the heavens for six days before a seventh day of rest. On Mount Sinai, God used this evidence to justify the Creation Sabbath law as a moral commandment (Exodus 20:8–11) to challenge skeptics who later would proclaim the six days of creation in Genesis a myth. This is the historical origin of the seven-day weekly cycle that the world currently follows. Only the weekly cycle (not months or years) is presented as a moral commandment because it is linked to creation. Nations have adopted weeks of varying numbers of days, but all have reverted to the seven-day week that God enjoins.[14]

Basic truths in religion may be expressed in parables subject to interpretation or as binding moral laws. The story about creation in Genesis 1 and the Creation Sabbath law in Exodus 20 are consistent. The story of creation reveals that the basic life forms were created differently within a matter of days by divine understanding with the mandate to reproduce after their kind, implying that there are no transitional stages linking the various species. This account is consistent with evidence in the living world and also in the fossil records at the Burgess Shale site in the Canadian Rockies and at the Chengjiang site in China, the only two deposits in the world where various forms of organisms are found together.[15] Because the scientific evidence (such as, DNA, fine tuning constants, abrupt gaps in the living world and fossil record) is consistent with the biblical creation narrative, therefore, on the subject of origins, religion should be our guide in interpreting the scientific data. For instance, according to empirical science, life cannot originate from nonlife (such as an inanimate bang) but only from a living being. Based on the Creation Sabbath law stipulated in the Decalogue we can only truthfully identify this being as the Abrahamic God.

The literalness of the six days of creation is the nucleus of the evolutionism-creationism controversy. Statements in the Bible such as "With the Lord a day is like a thousand years, and a thousand years are like a day" (2 Peter 3:8) are figurative. Otherwise, quantitative claims must be accepted as literal. God affirmed the literalness of the six days of creation in five ways. First, God specified that time period in a speech before a live audience. Second, God twice recorded the claim on tablets of stone: the first time is reported in Exodus 31:18 and 32:15–16, 19; the second time is described in Exodus 34:1–4, 28. Third, for the desert generation of Israelites, violation of the Creation Sabbath commandment was punishable by death (Exodus. 31:12–17; Numbers. 15:32–36). Fourth, God affirmed the literalness of the six days of creation by practical example with timed interventions to the Israelites during their forty years in the wilderness before arriving in Canaan. God specifically paralleled the six days of Creation with a six-day schedule of providing manna to feed the Israelites, after which they were commanded to observe a day of rest. Fifth, in addition to the Creation Sabbath law, God instituted a separate Sabbath related to agriculture. According to the Scriptures:

> The Lord said to Moses on Mount Sinai, Speak to the Israelites, and say to them: When you enter the land I am going to give you, the land itself must observe a Sabbath to the Lord. For six years sow your fields, and for six years prune your vineyards and gather their crops. But in the seventh year the land is to have a Sabbath of rest, a Sabbath to the Lord. Do not sow your fields or prune your vineyards.... (Leviticus 25:1–7, NIV)

> For six years you are to sow your fields and harvest the crops, but during the seventh year let the land lie unplowed and unused. Then the poor among your people may get food from it; and the wild animals may eat what they leave. Do the same with your vineyard and your olive grove. Six days do your work, but on the seventh day do not work, so that your ox and your donkey may rest, and the slave born in your house, and the alien as well, may be refreshed. (Exod. 23:10–12, NIV)

Although, the *seventh day* Sabbath of rest (which relates to the *origin* of the world) that people are to observe and the *seventh year* Sabbath that the land is to observe are both described as "Sabbath to the Lord," only the former is presented as a moral obligation in the Decalogue. The exclusive Jewish sabbaths and seventh year Sabbath to the Lord that do not relate to creation of the world are excluded in God's Ten Commandments. While six days is linked to creation of the living world, six years isn't. God's six days of making the earth and its surroundings habitable is not linked to the age of the earth. God could have accomplished this task in millions of years in which case the Sabbath separating the beginning and end of six days of creation would be irrelevant. God has the ability to create the world in a flash but this also would have no cosmological meaning. In the scriptures, the number seven is unique to God. The science of six days of creation, uniqueness of the seventh or Sabbath day, the origin of life, and the age of the physical world belongs exclusively to God. Different cultures may have different New Year but the seven-day week as a creation ordinance or edict is the same for all cultures suggesting the same God is responsible for the creation of all cultures. Scientists and clerics should stay within their disciplinary limits and humbly acknowledge that the immortal, invisible and illimitable God is the author of life and embodiment of their professions; without Jehovah/Yahweh/Allah/God, there will be no life, no science and religion.

In Exodus 20:8–11 concerning the Creation Sabbath law, God rebukes those who do not believe in the six days of creation presented in Genesis 1. Although the Bible is not a treatise on science, clergy today must not lose track of Albert Einstein's preposition that science without religion is lame. The unbiased clergy follows the Almighty God of Abraham, Isaac, and Jacob whose six days of creation are literal in meaning. The faithful believer does not venerate the modern god of evolutionism, which some liberal clerics today naively recommend to their congregations.

Truth #6: The absence of transitional stages in the fossil record or living world.

Uniformitarianism is the assumption that the same natural laws and processes that operate in the universe now have always operated in the universe in the past and apply everywhere in the universe.[16] Truth does not change with time, and in natural science, it is maintained by *fixed laws* and not by *random selection* or *theories* that are subject to change with the emergence of new evidence. Scientific claims or proofs must be observable directly or indirectly, testable, and repeatable. If any of these conditions is violated, a theory cannot constitute a foundational truth in science.

British biologist L. Harrison Matthews, in his introduction to a 1971 edition of Darwin's *The Origin of Species,* asserts:

> In accepting evolution[ism] as a fact, how many biologists pause to reflect that science is built upon theories that have been proved by experiment to be correct, or remember that the theory of animal evolution has never been thus proved?... The fact [theory] of evolution is the backbone of biology, and biology is thus in the peculiar position of being a science founded on an unproved theory—is it then a science or a faith? Belief in the theory of evolution is thus exactly parallel to belief in special creation—both are concepts which believers know to be true but neither, up to the present, has been capable of proof.[17]

The truth of transitional bacteria-to-human stages of evolution can only be scientifically established if they are observable, testable, and repeatable. The theory of evolution does not satisfy any of these conditions. World-renowned evolutionist Theodosius Dobzhansky affirms:

> These evolutionary happenings are unique, unrepeatable, and irreversible. It is as impossible to turn a land vertebrate into a fish as it is to effect the reverse transformation. The applicability of the experimental method to the study of such unique historical processes is severely restricted before all else by the time intervals involved, which far exceed the lifetime of any human experimenter. And yet it is just such impossibility that is demanded by

anti-evolutionists when they ask for "proofs" of evolution....[18]

The theory of evolution therefore falls short of classification as scientific truth. Why, then, is it fraudulently presented as if all scientific achievements in the world depend on it? The only logical answer is that evolutionism and creationism are both religious beliefs couched in scientific language.[19]

Evidence from the fossil record that should support evolutionary theory is circumstantial and not a proof of either evolution or creation. It nonetheless is relevant. Evolutionists and creationists reach different conclusions from examination of the same fossil data. The creationists claim that "The fossils say no," and the evolutionists respond that "The fossils say yes." The bottom line is that the truth about transitional stages in the formation of life must be consistent regardless of whether the evidence is from the fossil record or the living world. This fact is affirmed by deposits such as the ones described below in which there are no transitional organisms, just as there are none in the living world.

> The Cambrian fossils of Chengjiang in Yunnan Province, China (~525 million years ago), and the somewhat younger Burgess Shale fauna of Field, British Columbia (~515 million years ago) afford a magnificent diversity of organisms that represent all the major phyla in existence today, and some organisms that are as yet of enigmatic origin. This rapid change, a mere blink of an eye in geological terms, is what is called the "Cambrian Explosion."... Unlike many fossil deposits, those of Chengjiang and the younger Burgess Shale show exceptional preservation of non-mineralized body parts and internal soft tissues, affording us far more information than could be gleaned from preservation of hard skeletal elements alone.[20]

> These fossils are considered one of the most important fossil finds of the 20th century. Not only do they contain an exquisite degree of detail, but they also cover a diverse range of fauna, and are of significance in attempts to understand the evolution of life on Earth.[21]

> [T]he fossil record seems to show that most of the major animal groups appeared simultaneously. In the "Cambrian Explosion" we find segmented worms, velvet worms, starfish and their allies, mollusks (bivalves, snails, squid, and their relatives), sponges, brachiopods, and other shelled animals appearing all at once, with their basic organisation, organ systems, and sensory mechanisms already operational. We do not find crude prototypes of, say, starfish or trilobites. Moreover, we do not find the common ancestors of these groups.[22]

These findings are consistent with the creation worldview according to which the various organisms were created fully and simultaneously formed. The absence of common ancestors in such fossil deposits is consistent with evidence in the living world.

Evolutionists still have no explanation for these indisputable findings, as Coyne remarks in *Why Evolution Is True*.[23] Many evolutionists ignore all possible interpretations of this data from the Cambrian Explosion and prefer to base their conclusions instead on data from isolated deposits that are subject to equivocal interpretation. The truth nevertheless is expressed honestly by other evolutionists. For instance, Cambridge University botanist and evolutionist E.J.H. Corner posits, "Much evidence can be adduced in favour of the theory of evolution—from biology, bio-geography and palaeontology— but I still think that, to the unprejudiced, the fossil record of plants is in favour of special creation."[24] Oxford University zoologist and evolutionist

Mark Ridley agrees: "[N]o real evolutionist, whether gradualist or punctuationist, uses the fossil record as evidence in favour of the theory of evolution as opposed to special creation."[25] To insist otherwise is nothing but academic delusion. Darwin relied heavily on the fossil record to justify his theory, but as with what we observe in the living world it exhibits the absence of transitional organisms.

According to the Origin of Life Science Foundation, "Appealing to unknown laws constitutes a 'naturalism of the gaps,' corresponding to supernaturalists' appealing to a 'God of the gaps' for scientific explanation. Neither is acceptable in naturalistic science."[26] Similarly, appealing to unknown transitional organisms in the past in order to sustain the theory of evolution constitutes a "naturalism of the gaps" and cannot be part of true science let alone its foundation.

Truth #7: No DNA is evolutionary "junk."

DNA is central in evolutionary biology. Jerry A. Coyne's theory of "junk" DNA led him to conclude that evolutionism is true. Coyne's seductive explanation of dead genes as evidence of God's absence in the design of life forms prompted many distinguished evolutionists to praise his book titled *Why Evolution Is True.*[27] For example, Richard Dawkins, author of *The God Delusion* (2006), commented: "I once wrote that anybody who didn't believe in evolution must be stupid, insane or ignorant, and I was then careful to add that ignorance is no crime. I should now update my statement. Anybody who doesn't believe in evolution is stupid, insane or hasn't read Jerry Coyne. I defy any reasonable person to read this marvelous book and still take seriously the inanity that is intelligent design 'theory' or its country cousin, young-Earth creationism." The late Christopher Hitchens, author of *God Is Not Great* (2007), wrote: "Its ignorant opponents like to say that the process of evolution is 'only a theory.' (That's how they prove their ignorance.) Jerry Coyne shows with elegance and rigor that it is a hypothesis that meets and withstands all tests, and strengthens itself as a theory thereby. [One] could almost say that it had the distinct merit of being true." Steven Pinker, author of *The Stuff of Thought* (2007), stated: "Scientists don't use the word 'true,' but in this lively and engrossing book Jerry Coyne shows why biologists are happy to use it when it comes to evolution. Evolution is 'true' not because the experts say it is, nor because some worldview demands it, but because the evidence overwhelmingly supports it."

Arguing that dead genes are the ultimate proof of common ancestors in which those genes were once active, Coyne asserted:

> DNA sequencing supports the evolutionary relationships of species originally deduced from the fossil record. And, as natural selection predicts, we find no species with adaptations that benefit only a different species. We do find dead genes and vestigial organs, incomprehensible under the idea of special creation. Despite a million chances to be wrong, evolution always comes up right. That is as close as we can get to a scientific truth.... At this point I could simply say, "I've given the evidence, and it shows that evolution is true. Q.E.D."[28]

Coyne's book was published in 2008, but in the September 2013 issue of *Nature*, his theory was overturned.[29, 30] Once a deduction from DNA sequencing is proven wrong, it follows that the supporting deductions from the fossil record are also wrong, and so are some of the evolutionary predictions. Clearly evolutionary deductions do not always come up right millions of times as Coyne wants us to believe. Does this mean that evolution is not true? Not at all. Evolution is a secondary natural process that involves limited changes within the species boundary. The title of Coyne's book thus should have read, *Why Most Scientists Believe Evolutionism Is True.*

Coyne's book contains good science that shows his mastery in the field of evolutionary science, but the problem lies in his philosophical interpretation of the scientific evidence. His conviction is that life forms in our world are not perfectly designed and hence must be the products of random chance and evolution by natural selection. In real life, however, things that are intelligently designed by scientists are also subject to changes and corruption. *It is certainly a delusion to conclude that non-intelligence is the reason for imperfection.* The Holy Scriptures maintain that things were good at the time of creation, but were later subject to corruption as a result of human transgression. Evolution is thus not the only reason why things change over time. The Bible's overall trajectory indicates that things are deteriorating, and this is consistent with data in the natural world.

Because evolutionism is essentially a religion, evolutionists cannot accept that they are wrong. For instance, when Dawkins in a BBC-sponsored debate with Britain's chief rabbi, Lord Jonathan Sacks, was confronted with the ENDORE research finding that discredited the evolution theory of "junk" DNA, he remarked, "[I]t's exactly what a Darwinist would hope for, to find usefulness in the living world."[31] His answer contradicted views expressed in his book, *The Greatest Show on Earth: The Evidence for Evolution* (2009), published only two years before the ENDORE results. He there wrote:

> "Pseudogenes" are neutral for one kind of reason. They are genes that once did something useful but have now been sidelined and were never transcribed or translated. They might as well not exist, as far as animals' welfare is concerned. But as far as the scientist is concerned they very much exist, and they are exactly what we need for an evolutionary clock.... What pseudogenes are useful for is embarrassing creationists. It stretches even their creative ingenuity to make a convincing reason why an intelligent designer should have created a pseudogene—a gene that does absolutely nothing and gives every appearance of being a superannuated

version of a gene that used to do something—unless
he (God) was deliberately setting out to fool us.[32]

Perhaps pseudogenes are not exactly what scientists need for
an evolutionary clock after all. As expected, Dawkins book received
numerous (twenty-one to be exact) praises from evolutionists. PZ Myers,
Pharyngula's comment is noteworthy. He wrote: "A grenade thrown
right at the heart of the creationists. A fusillade of evidence backed
up by sound theoretical explanation. Beautifully explained. Will help
us build the intellectual foundation and the network of well-versed
literate elites who can address the rot at the root. Read it, please, please,
please."[33] Thank God, the grenade is rather shattering the evolutionary
foundation. If honesty is the rule, the new findings should dissolve
the latitude of these undue praises. What the public deserves is not
beautiful just-so stories; we need to see the transitional organisms in
our midst for science to remain true and timeless. Evolutionists who
boast about their ability to predict things have failed in this respect and
are not humble enough to admit it. It is better to make no prediction
at all than to make the wrong one for partisan purposes.

If the objective of the Clergy Letter Project is to bridge science
and religion, it is not an act of good faith to support the modern
scientific culture that insists, "Even if all data point to an intelligent
designer, such a hypothesis is excluded from science because it is not
naturalistic."[34] Instead, the logical way to promote a cordial interaction
between science and religion is embodied in the 1864 manifesto titled
"The Declaration of Students of the Natural and Physical Sciences,"
which was signed by 717 scientists including 86 Fellows of the Royal
Society:

> We believe that it is the duty of every Scientific
> Student to investigate nature simply for the purpose
> of elucidating truth, and that if he finds that some
> of his results appear to be in contradiction to the
> Written Word, or rather to his own *interpretations*

of it, which may be erroneous, he should not presumptuously affirm that his own conclusions must be right, and the statements of Scripture wrong; rather, leave the two side by side till it shall please God to allow us to see the manner in which they may be reconciled; and, instead of insisting upon the seeming differences between Science and the Scriptures, it would be as well to rest in faith upon the points in which they agree.[35]

To do otherwise is to pick and choose based on preference, a course that closes the pathway to truth and opens the door to endless controversy. The next chapter explains why Evolution(ism) cannot be a foundational scientific truth.

3

Why Evolution(ism) Is Not A Foundational Scientific Truth

Why Neo-Darwinism (the modern version of Darwin's theory of evolution, also called the Modern Synthesis) was the Biggest Mistake in the History of Science. The Modern Synthesis, while undoubtedly productive for a time, is a misconception of reality that has reached the limits of its explanatory power. The problems are fundamental. No amount of cosmetic surgery is going to correct them.[1]

—James MacAllister

The theory of evolution is not as scientifically sound as many people believe. In particular, the problem of the origin of life is well stated by the question, "Which came first, the chicken or the egg?" Every egg anyone has ever seen was laid by a chicken and every chicken was hatched from an egg. Hence the first chicken or first egg which appeared on the scene in any other way would be unnatural, to say the least. The natural laws under which scientists work are adequate for explaining how the world functions, but are inadequate to explain its origin, just as the tools which service an automobile are inadequate for its manufacture.[2]

—Theo Agard

Darwin pointed the direction to a thoroughly naturalistic—indeed a thoroughly atheistic—theory of phenotype formation; but he didn't see how to get the whole way there. He killed off God, if you like, but Mother Nature and other pseudo-agents got away scot-free. We think it's now time to get rid of them too.[3]

Atheist Jerry Fodor and Atheist Massimo Piattelli-Palmarini

Evolution or a change through modification (microevolution) or transformation (macroevolution) depends on an initial state or previous event; hence, by its dependence, a theory based exclusively on evolution cannot constitute a *foundational* truth in origin science. Whereas microevolution (changes within the species boundary), which depends on microcreation for its raw product, is a scientific fact, macroevolution (changes across the species boundary), which like macrocreation is not testable, is an indoctrinating belief that should also not be part of the science curriculum.

A foundational truth does not require millions of years to materialize because it is timeless. In the creation-evolution controversy the unequivocal

scientific and experiential truth in the living world
is the absence of the various transformational stages
predicted by macroevolutionary biology.

—Michael Ebifegha

Scientific facts are physical realities that cannot be the subject of
beliefs or debates. Scientific truths are derivatives of scientific
facts. Darwinian transformational evolution, such as from a land-
based vertebrate into a fish, assumes a sequence of events that cannot
be tested or repeated and hence cannot be classified as foundational
scientific truth. Deprived of the normal requirements that characterize
all scientific theories, Darwin's theory of evolution is based on analogy
or similarity arguments:

> **Analogy** would lead me one step further, namely, to
> the belief that all animals and plants have descended
> from some one prototype. But **analogy** may be a
> deceitful guide. Nevertheless all living things have
> much in common, in their chemical composition,
> their germinal vesicles, their cellular structure, and
> their laws of growth and reproduction.... Therefore
> I should infer from **analogy** that probably all the
> organic beings which have ever lived on this earth
> have descended from some one primordial form,
> into which life was first breathed.[4] (Emphasis mine.)

Notice that Darwin's key words are *analogy, deceitful, probably,* and
belief. He predicted the discovery of numerous transitional organisms in
fossil deposits, but the Burgess Shale and Chengjiang sites provided no
such evidence, confirming his own doubts about analogy as a deceitful
guide. Appraising his specious theory, he remarked:

> The main cause, however, of innumerable
> intermediate links not now occurring everywhere

throughout nature depends on the very process of natural selection, through which new varieties continually take the places of and exterminate their parent-forms. But just in proportion as this process of extermination has acted on an enormous scale, so must the number of intermediate varieties, which have formerly existed on the earth, be truly enormous. Why then is not every geological formation and every stratum full of such intermediate links? Geology assuredly does not reveal any such finely graduated organic chain; and this, perhaps, is the most obvious and gravest objection which can be urged against my theory.[5]

Darwinians usually do not admit that Darwin's theory can be wrong because the theory is justified by authority or expert opinion rather than hard scientific evidence or facts. The mistakes of Darwin's theory do not seem far-fetched. John Graham, in his response to the Royal Society's 2016 Evolution Meeting, wrote, "Evolution demands more transitions than extant (surviving or living) species, but the opposite is evident."[6] In the face of mounting empirical evidence that contradicts the status quo, evolutionists are inclined to incorporate empirical evidence as an extension or a revision rather than a substitute for the mistake or status quo. Challenging fellow evolutionists to embrace and absorb the truth, John Hands, author of *Cosmosapiens: Human Evolution from the Origin of the Universe*, made the following revolutionary comments in his presentation at the London 2016 evolution conference at the Royal Society:

> It's appropriate that this meeting is being held at the Royal Society, whose motto, we were reminded yesterday, is "Nullius in verba": Accept nothing on authority. The current paradigm in evolutionary biology, Neo-Darwinism, also called

the Modern Evolutionary Synthesis, has been the authority for some sixty-five years.... What we have heard over the last two days is empirical evidence that new species arise **rapidly,** from such mechanisms as symbiogenesis, horizontal DNA transfer, hybridisation, whole genome duplication, interactive systems producing novel emergent properties, and other mechanisms described in Part 2 of my book.

These mechanisms contradict the fundamental tenets of neo-Darwinism, namely:

- random gene mutations provide phenotypical characteristics enabling successful Darwinian competition;
- these random gene mutations spread through a population's gene pool by sexual reproduction;
- Darwinian gradualism leads to the genetic transformation of populations of individual species members over tens of thousands of generations;
- information flows one-way from a gene to a protein in a cell.

Not one whit of empirical evidence shows that new species arise from the neo-Darwinian mechanism.

To the contrary, Darwinian competition causes not the evolution of species but the destruction of species. It is collaboration in its various forms that causes biological evolution. Hence I'm surprised by calls for extending the neo-Darwinian Evolutionary Synthesis. You can't extend something that is broken. Surely what is needed now, after sixty-five years, is using the empirical evidence to develop a new paradigm for biological evolution.[7] [Emphasis mine]

While the tenets of Neo-Darwinism that is unable to form new species invokes *competition* and *gradual change*, the paradigm of Anti-Darwinism that produces new species advocates *collaboration/cooperation* and *rapid change*. Clearly, this justifies James MacAllister's use of "Why Neo-Darwinism Was the Biggest Mistake in the History of Science" in his review of the London 2016 Royal Society Conference. The empirical data are apparently consistent with the creation narratives in Genesis, in which the various kinds of species were formed rapidly and without competitions among the creatures or from other agencies. The fact that new species are produced quickly paralyzes Darwinism in three different ways. First, events that demand millions of years under the Darwinian paradigm will now require much shorter time intervals. This will require a complete revision of the evolutionary time scale. Second, intermediate or missing links are out of the question because there is just not enough time to establish the numerous transitional stages predicted by the current theory of evolution. This means the chapters on intermediate varieties or missing links in textbooks will be revised or disregarded. It is noteworthy that the Anti-Darwinism prediction (no missing links) is consistent with the fossil evidence at the Burgess Shale and Chengjiang fossil sites. The fossils at these sites are characterized as follows: they represent rapid change and cover a diverse range of fauna[8]. Third, with no intermediate or missing links, there is no common ancestor of all life on earth. The absence of common ancestors is consistent with what we observe at the Burgess Shale and Chengjiang fossil sites and in the living world; and it also confirms the predictions of the creation worldview. This shows that Darwin's tree of life is pure fiction and affirms *New Scientist*'s cover story, "Darwin Was Wrong: Cutting Down the Tree of Life" which was published on the eve of Darwin's bicentenary (24-30 January 2009 issue). This means a new theory of evolution is inevitable.

Perry Marshall, author of *Evolution 2.0: Breaking the Deadlock Between Darwin and Design*, in his review, "Royal Society's 'New Trends in Biological Evolution'—A Bloodless Revolution," also highlighted the historical significance of the 2016 Royal Society Conference. The following are excerpts of his report posted on November 30, 2016.

> In London from 7–9 November 2016 I witnessed a groundbreaking summit at the British Royal Society. Three hundred scientists from around the world gathered to evaluate a sea change in evolutionary theory.... This meeting had no mainstream precedent. Such a conference would NEVER have happened five years ago. It would have been too politically incorrect, too threatening to the Neo-Darwinian monopoly. . . .

> Until November 7–9, though, evolutionary theory was caught in the strangle hold of traditional evolutionary theorists. They have insisted for decades that chance and selection are the central driving forces of evolution....

> Experiments by Lynn Margulis, Eva Jablonka, David Prescott and Mae-Won Ho definitively prove: Not only do cells perform adaptations of astonishing sophistication in real time, these events are emphatically non-random.... This demolishes creationist/ID arguments that macroevolution is impossible. It's not only possible; *you can witness it in real time.... Empirical data also* demolishes the Neo-Darwinian doctrine that evolution is an aimless meander through random space.

> Neo-Darwinists permit no place for purposeful adaptations in their materialistic view. But now reductionism for the first time has been formerly challenged. The toothpaste is out of the tube and it

is not going back. There will be many more meetings like this. This was only the first....

Evolution going forward will not follow in the footsteps of its mannerless evangelists like Dawkins and Coyne. Conduct will be gentlemanly and respectful from now on. . . .

Scientism and reductionism have been punched in the face. Empiricism is making a comeback. Looking across the remaining years of the twenty-first century, the impact is difficult to estimate. But it will be great. "An era can be said to end when its basic illusions are exhausted."[9] [Emphasis his]

It is evident from both reports that this meeting signalled the beginning of the end of the Darwinian paradigm of evolution. Although the empirical data demolish the Neo-Darwinian doctrine that evolution is random and purposeless, does it also demolish arguments that macroevolution is impossible? The answer relies on subjective analyses because evolution is not an event that can be quantified. Evolution connotes change with time. Supposing micro (or small changes) and macro (or large changes) can both be accomplished within a short interval, then the description of these events becomes ambiguous and misleading if evolution is deemed to progress quickly from simple to complex organization or structures. To clarify the level of evolution involved in these empirical studies, let us consider a basic practical scenario as an illustration. For species X to be transformed into a different kind of species, Y, the former (X) will first be modified within its boundary. The changes that result in the modification process may be small (micro), such as bacteria developing resistance, or may be large (macro), such as bacteria developing tails, but they remain bacteria. Transformation into Y, which is the Darwinian paradigm of biological macroevolution (I prefer the term "extra-evolution," which means evolution resulting in transformation away from the species boundary into a different kind of species.), only happens when X can no longer be referred to as "bacteria" because it has been transformed morphologically (bodily) and habitually (mindedly) into a different kind of species, Y, that bears a different name, such as when fish are transformed into land vertebrates or early reptiles into birds. To date, no empirical data have shown a bacterium's being transformed into anything else but a bacterium, so we are still in the realm of the Darwinian paradigm of biological microevolution. (I prefer the term "intra-evolution," which means evolution resulting in modifications within the species boundary.) The Darwinian paradigm of biological macroevolution, or simply extra-evolution, theoretically demands millions of years, and so far empirical data disagree because it is pure fantasy. The absence of transitional stages in real life is proof that the Darwinian paradigm of biological macroevolution, or simply extra-evolution, is unattainable. If, however, we assume that the Darwinian paradigm of biological macroevolution has been achieved empirically and rapidly, then biologists face another hurdle. This means that Darwin's millions of years of evolution from bacteria to human, driven predominantly by natural selection, can be achieved instantly by a superior intelligent agency (God), a conclusion evolutionists detest, and this is why they are rallying clergies to support their theory.

The crucial issue is not whether the empirical data relate to microevolutionary or macroevolutionary events; the fact is that the Darwinian theory of evolution, based on random and gradual competition driven by natural selection, cannot produce new species in the empirical domain; new species are produced empirically and rapidly through collaboration or cooperation. The resulting controversy is not between creationists and evolutionists but within the evolutionists' camp. Previous efforts to point out the insufficiency of the theory of natural selection have been marginalized. Philosopher Fodor and biophysicist and molecular biologist Piattelli-Palmarini, for instance, in *What Darwin Got Wrong* (2010), made the following ground-breaking remarks:

> We close these prefatory comments with a brief
> homily: we've been told by more than one of our
> colleagues that, even if Darwin was substantially
> wrong to claim that natural selection is the
> mechanism of evolution, nonetheless we shouldn't
> say so. Not, anyhow, in public. To do that is, however
> inadvertently, to align oneself with the Forces of
> Darkness, whose goal it is to bring Science into
> disrepute. Well, we don't agree. We think the way
> to discomfort the Forces of Darkness is to follow
> the arguments wherever they may lead, spreading
> such light as one can in the course of doing so.
> What makes the Forces of Darkness dark is that
> they aren't willing to do that. What makes Science
> scientific is that it is.[10]

Fodor and Piattelli-Palmarini described themselves as card-carrying atheists and made it abundantly clear that their book was not about God, intelligent design, or creationism. They published in 2010, and the Royal Society's "New Trends in Biological Evolution" conference happened in 2016. In 2020, it is still business as usual in the scientific community, with no definite commitment to either extend or abolish the modern synthesis. No progress is to be expected as long as the old-school Darwinists and the New Atheists remain the dominant voice and authority within the scientific establishment.

What are the implications of the proceedings from this historic London conference? First, if evolution is not random or aimless, then it has a motive, purpose, and goal. Endowed with these capacities, evolution is indistinguishable from creation. Second, based on empirical data, natural selection is not the agent of evolution and hence loses its Darwinian glory as the only game in town. Third, the current theory of evolution is not a foundational scientific truth; a new theory of evolution based on empirical evidence is needed. Fourth, evolution cannot be a universal concept because empirically it is confined to intraevolution or changes within the species boundary. Fifth, human beings did not evolve gradually over millions of years from a common primordial ancestor. Sixth, dinosaurs did not become extinct millions of years ago, and this justifies why soft tissue and blood vessels have been found in some, others have been found pregnant, and some have dined on grass, contradicting the evolutionary time line or scale. Seventh, the economic impact of replacing Darwin's theory with a new theory of evolution will be enormous because new books on biological evolution will be required at all levels of learning. Eighth, the creationist and evolutionist designations within the scientific establishment will dissolve as empirical evidence prevails over authority and philosophical preference. Ninth, the science curriculum will be free of biases and philosophical preferences, and there will be no more court cases and no need for a Darwin Day or Evolution/Creation Sunday/Weekend. Tenth, scientism will be abandoned as scientists follow the evidence wherever it leads.

The preceding events are proof that just-so stories of the distant past, which no one witnessed, belong to the genre of faith and not science. The cell theory is a foundational scientific theory because it is based on empirical evidence. British biologist L. Harrison Matthews, in his introduction to a 1971 edition of *The Origin of Species*, reminded the world that Darwinian evolution, like creationism, is based on belief and as such cannot constitute a foundational scientific truth. Because Darwin's theory of evolution lacks empirical foundation, evolutionists in the academic institutions and media are constantly revising or extending it to embrace the collection of empirical data arising from advances in the field. Darwinists are selective regarding which empirical evidence they embrace and publicize because their goal is to prioritize the Darwinian paradigm. As an illustration, John Horgan is the director of the Center for Science Writings at the Stevens Institute of Technology. To rebuff challenges to the Darwinian theory of evolution originating from empirical data, he wrote the article, "Was Darwin Wrong? A Journalist Recounts the Epic Story of Modern Challenges to Evolutionary Dogma."[11] His article was posted in 2019; if preference is not the guiding principle, one would expect him to include at least the most recent work or event in the field. He made references to the following: a 2000 paper in *Scientific American*, "Uprooting the Tree of Life," by W. Ford Doolittle; a 2009 *New Scientist* cover story, "Darwin Was Wrong"; and *What Darwin Got Wrong*, by philosopher Fodor and cognitive scientist Piattelli-Palmarini (which he dismissed as "fatally flawed"). He made no reference to the revolutionary 2016 Royal Society's conference where scholars presented empirical evidence to show that the Neo-Darwinian theory of evolution, which is based on authority, is wrong and that a new theory based on empirical evidence is required. Based on his book *The End of Science* and the selections he made, Horgan wrote:

> In *The End of Science* I contended that science will keep extending and tweaking its current paradigms, like evolution by natural selection and the big bang, but there won't be any more comparably profound "revelations or revolutions."... To answer the question posed in my headline: Nah. Far from being wrong, Darwin is as right as ever when it comes to his big idea, natural selection.[12]

One way to arrive at the above conclusion is to assume that the 2016 Royal Society's summit did not take place, which is not the case. The other way is to admit that the meeting did take place, but the need for a paradigm change never came up, and there was no challenge to the request for further extension of the current Darwinian paradigm. The mere fact that evolutionists could not decide on whether to extend or reject the modern synthesis is clear indication that the Darwinian paradigm is in serious trouble.

Here we have an excellent example of how evolutionists use their position (authority) to promote only evidence that favours Darwinism. Although horizontal gene transfer may not be sufficient evidence to provoke revolutionary change in biology, the current empirical evidence that new species are produced through cooperation and rapid change is obviously revolutionary, which is presumably why Horgan skipped any reference to this meeting. By avoiding any reference to empirical evidence that contradicts the theory of evolution, Darwinists endeavour to conceal evidence that they do not like.

Is a theory that lacks empirical foundation hard to fix? Because Darwinism is a pseudoscientific religion, the empirical evidence is absorbed either through revision or extension of the theory. Because new species are formed empirically, evolutionists must choose either competitions with gradual change (Neo-Darwinism) or cooperation with rapid change (Anti-Darwinism) to resolve the current evolutionary controversy. Revisions or extensions of the Darwinian orthodoxy are out of the question because the only radical solution, as evolutionists MacAllister and Hands have pointed out, is a replacement of the Darwinian paradigm. If empirical evidence and integrity are the architects of science, then *to protect science's integrity*, authorities, however eminent, cannot override empirical evidence.

Natural selection as the primary driving force of evolution can never select between Neo-Darwinism and Anti-Darwinism; otherwise, it will be deemed as conscious and mindful. Thus, contending that Darwin is as right as ever concerning the role of natural selection is both false and inflammatory. Atheistic evolutionists Fodor and Piattelli-Palmarini's proposition that something is quite possibly fatally wrong with the theory of natural selection cannot after all be dismissed as fatally flawed, as Horgan would want us to believe. By dismissing the honest views of other accomplished evolutionists as totally flawed, Darwinists attempt to suppress the truth and silence scholars who dare to question.

By asking judges and clerics to declare evolution as the only scientific worldview, evolutionists expose their vulnerability and use the opportunity to eliminate what they fear. Darwinists may resist comparing their worldview with the creation worldview in science classes; however, they cannot resist or ignore detailed comparison of their theory with empirical evidence procured by scientists. Why? Because the creation worldview without an empirical foundation is a story based on creation, and the evolution worldview without an empirical foundation is a story based on evolution, but empirical evidence in science is not a story. Hence empirical data are neutral and independent of any worldview; they can only be used to monitor the validity of the various worldviews. It is evident that the empirical evidence so far fits the creation model and contradicts the current evolution model.

The honest and logical conclusion is that Darwin's "great tree" with numerous branches, representing different species diverging from a common ancestor, cannot be achieved with natural selection as the dominant mechanism. The cat is out of the bag; it is time for science to regain its empirical right and shed its religious and political garments; it is time for honest scientists to reject allegiance to philosophical preferences and refrain from wrongful submission to authority in an effort to sustain Darwinism; as loyal servants of the state, they should faithfully follow the empirical evidence to wherever it leads and accurately report whatever they find. The public deserves nothing else but the truth.

Whether created or evolved, things within a system change with time. For instance, consider cell division (natural evidence of microcreation that produces a primary change in the biological system) and mutation (natural evidence of microevolution that produces a secondary change in the biological system). These are empirical facts, but the interpretation is subject to philosophical preference. The term "creation" as designating a natural process is controversial in science, where it invariably implies a creator. Creation, whether it occurs naturally or supernaturally, is currently perceived as a religious event. DNA replication, cell division, and protein synthesis are all natural processes of creation. These processes are incorrectly accredited to evolution. This is evidence that science does not actually know the difference between natural creation and evolution. If empirically evolution is inseparable from creation as natural processes, how would scientists integrate the two? For an authentic scientific formulation, wisdom demands that creation as a natural and primary process should be associated with origins and diversity among the basic kinds of species, while evolution as a natural and secondary process should be linked only with diversity within a given kind of species. The modern myth is that creation offers no predictions and is therefore unscientific, whereas evolution does and is. Although the capacity for prediction is important, what drives genuine scientific theory is understanding, critical observation, testability, repeatability, and the reproducibility of phenomena. These are prerequisites for a correct scientific theory. Prediction is not a prerequisite for the formulation of a scientific theory but instead a consequence or outcome (postrequisite) of an authentic scientific theory. Creationism and intelligent design are not classified as scientific theories because, like evolutionism, they lack the aforementioned prerequisites. Casey Luskin, in his article, "Design vs. Descent: A Contest of Predictions,"[13] compared predictions from both worldviews (see Table 3.1) to justify why the ability to predict should not be the central issue.

Table 3.1: A Comparison of Predictions from Intelligent Design and Descent with Transformation

Line of Evidence	Prediction from Descent	Prediction from Design	Data	Best Explaining Theory
1. Biochemical Complexity	High information content and machine-like, irreducibly complex structures will NOT be found.	High information content and machine-like, irreducibly complex structures will be found.	High information content and machine-like, irreducibly complex structures are commonly found.	Design
2. Fossil Record	Forms will appear in the fossil record as a gradual progression with transitional series.	Forms will appear in the fossil record suddenly and without any precursors.	Forms tend to appear in the fossil record suddenly and without any precursors.	Design

3. Distribution of Molecular and Morphological Characteristics	Genes and functional parts will reflect those inherited through ancestry and are only shared by related organisms.	Genes, DNA sequences, and functional parts will be reused in different unrelated organisms.	Genes and functional parts often are not distributed in a manner predicted by ancestry and are often found in clearly unrelated organisms.	Design
4. Biochemical Functionality	The genetic code will contain much discarded genetic baggage code or functionless "junk DNA."	The genetic code will NOT contain much discarded genetic baggage code or functionless "junk DNA."	Increased knowledge of genetics has created a strong trend toward functionality for "junk DNA." Examples of DNA of unknown function persist, but function may be expected or explained under a design paradigm.	Design

Once a theory lacks prerequisites, it cannot constitute a foundational scientific truth. This explains why several of the evolutionary predictions are wrong. Some of their philosophical conclusions are pure fictions (such as life from nonlife) because their goal is to eliminate any reference to God in the natural world. Affirming this point, evolutionist Richard Lewontin of Harvard University wrote in the *New York Review of Books* that materialism is absolute, so no Divine Foot is allowed in the door.[14] Evolutionist Scott C. Todd of Kansas State University also maintained that any data that points to an intelligent designer will be excluded from science because it is not naturalistic.[15] A theory that constitutes foundational truth must be universal and cannot be based on censored or partial data.

For the biased evolutionist, sticking to philosophical preference is the primary goal; the truth is secondary.

Consider the following scenario. Two astronomers, Mary, a creationist, and Jack, an evolutionist, find themselves on an Earth-like planet. Upon landing, they find two sets of information programs: Set A, which parallels the computer program on Earth, utilizes a binary code (programming database with a two-numeral system using the digits 0 and 1), and Set B, which parallels the naturally occurring DNA program on Earth, uses a quaternary code (programming database with a four-numeral system using the digits 0, 1, 2, and 3). In a news conference on Earth, the two astronomers provide a report of their findings to the press. Creationist Mary asserts that both sets of discoveries are information-processing systems, but Set B is a more sophisticated network. As a result, she concludes that the origins of both programs are to be credited to intelligent design. Evolutionist Jack maintains that both are programming networks, and Set A, by its configuration, is less advanced than Set B. However, on grounds of philosophical preference, he invokes the principle of Occam's razor to conclude that, although he accepts that humanlike intelligence is involved in the programming of Set A, he disagrees that a similar intelligence is involved in Set B because it resembles the naturally occurring DNA program found on Earth that evolutionists—to justify Darwinian evolution—attribute to unintelligent design. Therefore, because of the philosophical paradox, he attributes both to nonintelligent design and dismisses any alternative conclusion as unscientific.

Mary and Jack examined the same scientific evidence and provided similar scientific interpretations, but they differed in their conclusions on philosophical grounds. Which of these two scientists would be deemed both honest and unbiased in presenting the scientific truth? Objectively speaking, Mary is both honest and unbiased, while Jack is honest *but* biased. However, for many modern scientists who teach that naturally occurring DNA, with its four nitrogenous bases of adenine, cytosine, guanine, and thymine, is the product of unconscious and unintelligent design, Jack's analysis is the preferred choice because it is focused on authority and does not give credit to an unknown designer or deity.

Evolutionists, without first establishing the origin of life, claim to have addressed the origin of species, prompting some clerics to endorse evolutionism as foundational scientific truth. But one would expect a foundational scientific truth to address the origin of life and the origin of species simultaneously. The Episcopalian clergy accordingly maintains that the theory of evolution provides a fruitful, unifying, and scientific explanation for the emergence of life on Earth, which is far from the truth. In order to limit explanations to the Darwinian paradigm of evolution, a $10 million Evolution 2.0 Prize for Life's Origin is being offered to anyone who can produce a self-organizing digital communication system.[16] The expectation is that an explanation of how to get from chemicals to codes may be as revolutionary as Einstein's theory of relativity or the development of the transistor and would ultimately resolve the questions surrounding the origin of life. The primary difference between the current $10 million Life Origin award and a $1 million award by the Origin-of-Life Foundation is that the latter was kept a secret from the public, while entrepreneur Perry Marshall and investor Kevin Ham publicly announced the former at the Royal Society in Great Britain on May 31, 2019. The Origin-of-Life Foundation's contest was discontinued because no one was able to provide mechanistic answers to how "seemingly unintelligent natural processes could have written such highly prescriptive recipe/message linguistic-like code."[17] The world anticipates and welcomes a response to the $10 million prize that is expected to break the deadlock between Darwin and design. The issue is not about money but about literal truth that is outside the purview of pure science. The truth is that natural processes are lifeless and therefore cannot be the first cause of life because this would violate the scientific law of biogenesis and prohibit the principle of uniformitarianism in science. Creating a self-organizing digital network using abstract intelligence will be a great achievement, but this will only affirm the use of intelligence to uncover intelligent design at work and will certainly not be proof of life's origin.

Before we delve further into specifics, we must clarify some important points. The living human being is characterized by a material component (e.g., the brain) and an immaterial dimension (e.g., the mind). Analysis of the brain by use of the five senses (sight, hearing, touch, smell, taste) cannot be applied to the mind. The dead body, of course, has no mind; the brain remains but is nonfunctional. The scientific law of biogenesis stipulates that life can only originate from preexisting life. This law retains its usual meaning when the term "life" is replaced by "mind," but it is meaningless when the word "brain" is substituted. Fossils are mindless; therefore, analyses based on their record cannot correctly address the origin of species. However, because science focuses on the brain, whereas religion addresses the mind, a complete description of origins is possible only when the evidence from science *and* the revelations from religion are interpreted in light of each other, consistent with Albert Einstein's proposition that science without religion is lame and religion without science is blind. Philosophical conclusions based exclusively on scientific findings are, therefore, not scientific facts but opinions or beliefs. Acceptance of these conclusions is based on faith in "expert" opinions. Thus, origin studies involve two types of faith: faith in the claims of a Creator who masterminds both the material (natural) and immaterial (supernatural) realms, and faith in the claims of experts whose understandings are limited to the material realm.

The origin of life poses another problem. Scripture describes life as the breath of God. At death, the breath of God escapes from the body, which loses its vital and functional capacity. One must not conflate the breath that sustains normal life with the breath maintained by ventilators to prolong the life of a patient who would otherwise be declared dead. Because God is a Spirit, this implies that a veridical investigation of the life's origin must involve study of intact living systems from the physical and spiritual perspectives and not just chemical composition. Because the dead body has no life, it means life itself functions as an immaterial switch, and its origin is outside science's purview. Creating a self-organizing digital network will not resolve the origin-of-life controversy. The obvious conclusion is that evolutionary theory, which addresses only the material aspects of living systems, cannot constitute a foundational scientific truth on the origin of species. At best, it can address only the diversity of species.

Unable to suppress opposition from creationists within the scientific establishment, evolutionists have chosen to promote their views to the religious community through their clerics, who are still recovering from the mistakes of the past, including their predecessors' challenge of the credibility of Galileo's heliocentric solar system. For this reason, these clerics are prepared to bow to majority opinion within the scientific establishment. They fail to realize that all discarded scientific theories were once supported by a majority of its members. Benefiting from their vulnerability, evolutionists in the scientific community now see the clerics as their spokespeople in promoting their doctrine and challenging the views of colleagues who are diametrically opposed to their worldview on purely scientific grounds. In the following we will consider two captivating but simple arguments concerning Earth's age and dinosaur evidence that proevolution scientists present in defence of evolutionism as foundational science.

For a highly secularized world, the age of the earth is the litmus test for determining which worldview makes sense. One of the major reasons why evolutionism is endorsed is to avoid ridicule involving the plausible age of the earth. There is a widespread misconception that according to the Bible the earth is six thousand years old, but the Holy Scriptures say nothing of the kind. It was Archbishop James Ussher (1581–1656) of Ireland who posited that the world was created on October 23, 4004 B.C. based on the genealogies found in Genesis. But the age of the earth cannot depend on the ages of its inhabitants, just as the age of a building cannot be based on the ages of its occupants. The true age of the earth, like the origin of life, is indeterminate. For God, the time domain runs eternally from negative infinity to positive infinity. In cosmological terms, the time t = 0 in God's agenda marks the boundary between the immaterial (invisible) and the material (visible) world. A fair analogy to the immaterial-versus-material domain can be illustrated from a mathematical point of view. When we solve a quadratic equation such as $t^2 = 9$, where t is the time in seconds (s), we obtain answers suggesting that time is both positive (t = 3s) and negative (t = -3s). We accept the positive solution because it intersects with our material world and declare the negative solution inadmissible because it has no meaning, being outside our material world. In other words, the negative answer can be perceived as belonging to the immaterial or invisible domain.

The New Testament contends, "By faith we understand that the worlds were framed by the word of God, so that the things which are seen were not made of things which are visible" (Hebrews 11:3, NKJV). The Old Testament records, in the Genesis account of creation, that "in the beginning God created the heavens and earth" (Genesis 1:1). On the one hand, the immaterial or supernatural world has no beginning and no end; on the other hand, the material or natural world has a beginning and end. Recall Jesus Christ's remark to the Pharisees who challenged him: "You are from below; I am from above. You are of this world; I am not of this world" (John 8:23, NIV). Additionally, during his farewell prayer in the presence of his disciples, Christ made reference to his relationship with God *before* the world began (t = 0) in John 17:5. In responding to Job, God revealed that angels witnessed and rejoiced at the laying of the earth's foundation (Job 38:4–7). When the aforementioned scriptures are examined in context, they suggest that at t = 0 the physical heavens and earth were not made from materials of this world but from the invisible world.

The practical way to express this idea is that at the beginning of time, things in the immaterial domain, by God's creative power, wisdom, and understanding, became visible in the natural world as parameters of time, space, and matter. According to Genesis 1, over the span of six days God modified the raw, unordered earth to accommodate plant and animal life. In the Genesis account the heavens and earth of an unknown age appeared at t = 0; Ussher, in contrast, assumed in his analysis that at t = 0 the age of the earth was zero. Genesis 1:2, although without the phrase "And God said let there be the heavens and earth and the seas and it was so," unequivocally stipulates that the raw earth came and was "without form and void; and darkness was on the face of the deep; and the spirit of God was hovering over the face of the water." The fact that on the first day of creation light was called into existence in the physical world by God's command validates the view that the primordial earth of an unknown age, covered in water, was also ordered to emerge from the invisible into the visible domain. Empirical scientific analysis reported in Chapter 2 of this book (see reference 1) affirms the fact that Earth may have had water from Day One of creation, which contradicts the evolution worldview that water came much later. Though God commanded some things from the supernatural into the natural world to accommodate life, other things were made later from things already in existence in the natural domain. For instance, the sun, moon, and stars were made on the fourth day. From the preceding discussion it follows that we cannot determine the earth's age by either interpreting the scriptures literally or in any form with reference to t = 0 or by measuring the ages or properties of the elements it comprises. Next, we shall consider the degree of truth of the scientific estimates of the ages of the earth and the universe.

Age determination by scientific methods is a complex issue because of the different types of ages. Chronological age specifies the number of years we have been alive (twins, for example, have the same chronological age); biological age shows the rate at which we age (twins may have different biological ages, which is influenced by environmental, genetic, and other biological factors); geological age addresses the earth's history and records the ages of its layers and constituents. The structural age of the earth, conversely, measures how long the planet has been in existence and should be considered independently of the geological ages of its layers or constituents, just as the age of a building is independent of the ages of its constituents or components and occupants. Because science cannot exclusively measure the age of a building based on anything or by any means, it also cannot exclusively measure the structural age of the earth based on anything or by any means (including radioactive dating). For example, suppose there are two buildings that were built ten years ago (structural age); one is formed with radioactive substances in its matrix and the other without them. Scientific age determination will not be able to provide an estimate for the building without the radioactive substance and may estimate billions of years for the one with radioactive substance. Therefore, the measured value of 4.5 billion years as the earth's age is not the true structural age, which is unknowable for the reasons discussed in the following.

To determine how old a person is today, we utilize A.D. (Anno Domini, which in Latin means "in the year of the Lord" or, alternatively, "after Jesus Christ was born") as the established benchmark. That is, on the age axis, the origin (age = 0) is specified as 0 A.D. Thus, *2017 A.D.* means "2017 years after Jesus Christ was born." With A.D. as the known benchmark, a person's age becomes the difference between today's date and the person's date of birth. When scientists say that the earth is 4.5 billion years old, and the universe 15 billion years old, technically they are assuming an arbitrary benchmark that may fluctuate; this presumably explains why a recent study reported on May 18, 2019 claimed the universe to be a billion years younger,[18] while another study published on September 12, 2019 contained the statement that the universe might be two billion years younger, with the authors declaring that it seems to be looking younger every day.[19]

Evolutionism, by this logic, is predicated on an imaginary frame of temporal reference and hence cannot be foundational scientific truth. If, in contrast, proevolution scientists contend that this imaginary reference approximates the time t = 0 when the big bang occurred, we run into a different set of problems. The big bang theory contradicts the scientific law of biogenesis by insinuating that life originated as the result of that uncontrolled burst of concentrated matter and energy. To make sense, and account for the prevalence of natural laws and resources, life, and intelligence on Earth, it would be more feasible to contend that a controlled outburst of power generated time, space, matter, natural laws, and physical constants in the natural domain. The discovery of the God particle in physics did not happen by chance but was a controlled event, masterminded by high-level intelligence drawn from various universities around the globe.

To show that scientists are just as ignorant as everyone else about the earth's age, it is important to examine the history of their adventure. Before the late 1700s, scientists estimated the earth's age to be about six thousand years.[20] However, because they realized that geological processes such as erosion and sedimentation occur over extremely long periods of time, they perceived that the earth should be much older. Science thus faced significant problems from the start. First, there was no empirical benchmark for a six-thousand-year-old Earth. Second, how could erosion and sedimentation, or any natural process, increase the earth's age from six thousand to 4,500,000,000 years? A Google search asking "How did water come to Earth?" or "Where did Earth get its water?" reveals that scientists are still trying to figure out the answer. Of interest is a recent research finding by Dr. Lydia Hallis at the University of Hawaii:

> We cannot rule out the addition of water to Earth's surface after its formation (i.e., via comets and asteroids), but our data suggests that Earth had water from the very beginning of its formation, so a large amount of water addition later was not necessarily needed to produce our oceans. We can say that the water we measured from the deep mantle is highly unlikely to have been added in this way, because cometary and asteroidal impacts would not have been large enough or powerful enough to affect the deep mantle thousands of kilometers beneath the surface, and previously reported geochemical data suggests the source regions for our rocks have not been disturbed in about 4.5 billion years.[21]

The scriptures contend that in the beginning, God set the earth on its foundations (Psalm 104:5; Job 38:4). The earth was covered with the deep as with a garment, and God separated the waters to create the sky and expose dry land (Genesis 1:1–10; Psalm 104:6– 9). These scriptures imply that erosion and sedimentation had no role in forming the earth. The aforementioned scientific data from the University of Hawaii in 2015 suggests that Earth had water from the very beginning of its formation, eliminating the need for additional water from comets and asteroids to form the oceans, which is consistent with the biblical account stating that the earth and the seas were together from the beginning. In 2010, simulation studies by Nora de Leeuw's team at the University College of London[22] also suggested that the earth may have had water from day one, which further supports the idea that erosion and sedimentation are not responsible for Earth's estimated age of 4.5 billion years. In addition, in 2008, Australian scientists led by Dr. Nicolas Flament from the University of Sydney showed that early Earth was covered in water.[23] Moreover, the events following the eruption of Mount St. Helens in 1980 defied the theories of catastrophism and uniformitarianism formulated to support evolutionism. Within five years, the eruption generated strata that essentially duplicated what, according to evolutionists, would require thousands of years for slow and gradual processes to accomplish.[24] The reasonable conclusion is that the layering found in the earth's geological strata is not a function of erosion or sedimentation but, rather, a property of nature that has bypassed scientific scrutiny.

The truth is that the exact number of years the earth has existed is unknowable; the earth is neither young (six thousand years) nor old (4.5 billion years) because God renews its face (Psalm 104:30). Because neither science nor any other discipline can provide the true age of the earth, clerics, in an effort to be educationally, politically, or socially correct, should not be obliged to accept the opinions of scientists. Evolutionism is a paradigm that relies on very long ages, and if the age of the earth cannot be determined accurately, then it cannot constitute a foundational truth. Given the debunking of Earth's longevity as either six thousand or 4.5 billion years, evolutionists invariably invoke the evidence of dinosaurs.

The word "dinosaur," which literally means "terrible (*deinos*) lizard (*sauros*)," has long captivated people's attention. Because the word does not appear in the Bible, evolutionists, as the dominant voice of the modern scientific establishment, use the discovery as circumstantial evidence for advocating evolution(ism) as a foundational scientific truth on the subject of origins. For instance, Brian J. Alters and Sandra M. Alters in *Defending Evolution*, which was dedicated to instructors who teach evolution, wrote: "The student might be encouraged to consider that no scientist has ever directly observed a dinosaur—just fossils of 'supposed' dinosaurs.... So even though scientists did not observe living dinosaurs, their existence is nonetheless considered a scientific fact. Likewise, humans did not directly observe the evolution of dinosaurs, but their evolution is nonetheless considered to be scientific fact."[25] Had the Alters been creationists, they might very well have concluded, "Likewise, humans did not directly observe the creation of dinosaurs, but their creation is nonetheless considered to be scientific fact." To affirm how evolutionists regard the evidence of dinosaurs as pivotal, I include below a brief excerpt from a BlueInk review of my book *Creation or Evolution? Origin of Species in Light of Science's Limitations and Historical Records*:

> Ebifegha sets out to scientifically debunk Darwin's theory of evolution while championing biblical creationism.... He also criticizes evolutionists for preaching "science of the gaps," a tendency to assume that we can fill in crucial gaps in the fossil record and extrapolate the existence of transitional forms. Yet he believes that the Bible is an historical document from which we can interpret the history of living things, despite its own glaring gaps. It makes no mention of dinosaurs, for instance, nor does it

explain why God would create so many species only to watch them go extinct over the eons.[26]

The existence of dinosaurs, just like that of every other creature, is a fact of history and not a fact of either science or religion. In the Bible, the Book of Job describes such animals as bears, lions, ravens, mountain goats, wild asses and oxen, ostriches, horses, locusts, hawks, eagles, behemoths, cattle, leviathans, and fish (Job 38–41). In science the outstanding creatures of the past are the dinosaurs; in religion they are the gigantic behemoths and ferocious leviathans. The names given to these creatures in science and religion are different, but their descriptions are similar, and they all are extinct. Describing the mighty behemoth that existed at the same time as human beings, God said:

> Look now at the behemoth, which I made *along* with you; He eats grass like an ox. See now, his strength is in his hips, And his power *is* in his stomach muscles. He moves his tail like a cedar; the sinews of his thighs are tightly knit. His bones *are like* beams of bronze, His ribs like bars of iron. He is the first of the ways of God; Only He who made him can bring near his sword. Surely the mountains yield food for him, And all the beasts of the field play there. He lies under the lotus trees, In a covert of reeds and marsh. The lotus trees cover him *with* their shade; The willows by the brook surround him. Indeed the river may rage, *Yet* he is not disturbed; He is confident, though the Jordan gushes into his mouth, *Though* he takes it in his eyes, *Or* one pierces his nose with a snare. (Job 40:15–24, NKVJ)

Several Bible commentaries have referred to the behemoth as either an elephant or hippopotamus because they eat grass and are comparatively bigger than all other herbivores. Biblical comparisons with cedars normally suggest great height. For instance, Isaiah 3:13 says, "For all the cedars of Lebanon, tall and lofty, and all the oaks of Bashan" (NIV). When applied to the tail of an animal, the cedar simile should be in terms of length, but the tails of elephants (2.5–5 feet) and hippopotamuses (0.5–2 feet) are not especially notable. The appropriate reference for an elephant would be its large floppy ears or long trunk and for a hippopotamus its huge mouth or massive jaws. The description of the behemoth's tail that sways like a cedar and its strong limbs that are like rods of iron precludes any currently extant creatures. Instead it appears to be more representative of the gigantic sauropods that had massive tails of up to 56 feet in length.[27]

Evolutionists scoff at the Bible's possible reference to a grass-eating dinosaur because according to the geological timescale, dinosaurs became extinct 65 million years ago and the first grasses appeared 55 million years ago.[28] Sauropods flourished 208 to 146 million years ago,[29] so that it would be a miracle if any trace of grass were found in their digestive systems. Apparently, however, this miracle occurred in 2005, when it was discovered that the dinosaur *Titanosaurid sauropods* ate grass. Our Creator, the God of the Bible, was right that dinosaurs lived contemporaneously with people, thereby discrediting the evolutionists' arguments.[30] God's testimony was further justified in 2005 with the discovery of soft tissue, blood vessels, and cells in the thighbone of a tyrannosaurus fossil.[31] Mary H. Schweitzer, in her study of an older tyrannosaurus at the North Carolina Museum of Natural Science, was able to verify that it was pregnant.[32] She also examined an 80-million-year-old, plant-eating, duckbill dinosaur. In order to confront those who rebutted her findings, she sent specimens to different labs for independent tests, all of which confirmed her results that the structures observed were primeval soft tissues and not biofilm. Evolutionists believed that these structures would have disappeared over several thousand years. Because the evidence is a scientific fact, it means that the geological timescale is not a scientific fact but an arbitrary timescale to sustain the evolution worldview.

Let us now consider the leviathan, which scholars wrongly assume to be a crocodile. This creature's amazing structure and characteristics are outside science's expectations. Part of the Bible's description of the leviathan reads as follows:

> His sneezings flash forth light, And his eyes *are* like the eyelids of the morning. Out of his mouth go burning lights; Sparks of fire shoot out. Smoke goes out of his nostrils, As *from* a boiling pot and burning rushes. His breath kindles coals, And a flame goes out of his mouth…. He regards iron as straw, *And* bronze as rotten wood. He makes the deep boil like a pot; He makes the sea like a pot of ointment. (Job 41:18–31, NKJV)

Only those who saw the leviathan could testify to its uniqueness, as in Psalms 104:25–26: "There is the sea, vast and spacious, teeming with creatures beyond number—living things both large and small. There the ships go to and fro, and the leviathan, which you formed to frolic there" (NIV). The leviathan must have been special to be the only creature referred to by name in this text. One can only conclude that the Bible is not silent on stupendous creatures that are now extinct.

If evolutionism cannot be foundational scientific truth, what then is? Because no one observed primal creation, a prudent creator of this world would claim credit, but an irresponsible one would not. As previously discussed, the historical record shows that only the Judaeo-Christian God, in speech before a live audience of witnesses and in print on stone tablets, has claimed credit for having created the physical world in six days to accommodate life, with a seventh day of rest as a seal separating the beginning of the first day of creation from the end of the sixth day of creation.

As the Creator, God is the embodiment of true religion and science. In science, God speaks through nature, and in religion, God speaks through agents to enlighten people on how to live peacefully and righteously. In both disciplines, God governs through laws. In religion, the Sabbath law in the Decalogue links the origin of the material world to God; in science, the law of biogenesis links the origin of life to God. God is the greatest scientist and the reason for scientific resources. Sir Isaac Newton acknowledged this fact by recognizing God as personal and crediting all his scientific discoveries to God. Albert Einstein, although he did not believe in a personal God, nevertheless acknowledged this fact when he defined God as an illimitable spirit, a superior reasoning power that is revealed in the incomprehensible universe. Francis Collins, former director of the National Human Genome Research Institute, after converting from atheism to theism, described God as the greatest scientist and in his book, *The Language of God*, acknowledged the DNA assembly as the language in which God created life. God's revelation through science and religion is therefore the foundational truth.

In the face of God's personal claim of ownership of the universe— and hard scientific evidence backing this claim—there are those who, based on personal reasons and/or frustrating world events executed by religious fanatics, fraudulently use science to describe God as a delusion. The lead scientist in this regard is atheist and evolutionary biologist Richard Dawkins; distinguished scientist Stephen Hawking was a recent convert to atheism. To these and others, Einstein's statement applies: "In view of such harmony in the cosmos which I, with my limited human mind, am able to recognize, there are yet people who say there is no God. But what really makes me angry is that they quote me for support of such views."[33] To the new atheists of this age, Einstein's rebuke also applies: "Then there are the fanatical atheists whose intolerance is of the same kind as the intolerance of the religious fanatics and comes from the same source."[34] Their goal is to use their influence to undermine God's existence and seize the opportunity to promote evolutionism as foundational truth.

Today, honest evolutionists are abandoning the current neo-Darwinian theory of evolution, also called the modern synthesis, because of its failure to align with empirical data. Perry Marshall affirms this fact by referencing the current research undertaken by Denis Noble, a fellow of the Royal Society and former president of the International Union of Physiological Sciences. Below is an excerpt from Marshall's article, "83-Year-Old Heart Scientist Rocks the Foundations of Evolution":

> Denis Noble is the guy who figured out the cardiac rhythm which made pacemakers possible. He was the first person to model a human organ on a computer – in 1960.... Based purely on heart research, he concluded evolutionary biology had major, fundamental, foundational problems (picture building a four-story house on swampland that floods twice a year). This is because when Denis "knocked out" genes that regulate heart rhythms, what happened was entirely different than what "selfish gene" evolutionary theories predicted would happen. Denis concluded that the Modern Synthesis had literally gotten cause and effect backwards. By ignoring major aspects of systems biology, evolutionary biologists had shot themselves in the foot [in] two ways: First, they had exposed the field to relentless criticisms from people who doubt evolution itself. Second, medical and disease implications of evolutionary theory were being misunderstood and miscategorized. This has been a very expensive mistake, not just financially but in terms of lives.[35]

Evolutionists invariably maintain that, like every other well established scientific theory, no findings can overturn the accepted theory of evolution. In the next chapter, we will learn why both the pro-evolution and pro-creation Clergy Letters are not the appropriate solution to the ongoing evolutionism-creationism controversy.

4

Follies in the Evolution and Creation Clergy Letters

Clergy, unless they are scientifically trained, shouldn't be affirming "foundational scientific truths." I mean, we wouldn't want clergy affirming that mental depression is caused by demons, would we? Odd that the clergy who can affirm, as clergy, a "foundational scientific truth" end the letter by asking that science remains science and that religion remain religion.[1]

—Joel Watts

New Scientist magazine in 2009 asked sixteen of the world's most eminent evolutionary biologists to identify the biggest gaps remaining in evolutionary theory. Their responses published in the 31 January-6 February issue indicate that the prominent big questions they would most like to see answered are the origin of life itself and the dynamics of natural selection. To mention a few, Niles Eldredge desires to know the ecological context in which selection operates; Steven Pinker wants to know how selection leaves its fingerprints on the genome; and Eörs Szathmáry wonders how evolution by natural selection explains complex thought.[2]

These responses suggest that evolutionists are just as ignorant of the mechanisms that underpin their worldview as they are of the mechanisms that underpin the creation worldview. Little wonder that some of their major predictions such as "junk" DNA turn out to be wrong! These are serious gaps in the evolutionary theory that the pro-evolution Clergy Letter Project failed to incorporate in endorsing the theory of evolution as foundational scientific truth.

Here is the inconsistency. When scholars believe in creation and do not understand how it works, the paradigm is classified under religion. But when scholars believe in evolution by natural selection and do not understand how it works, the paradigm is still retained as a foundational scientific theory.

—Michael Ebifegha

This chapter is critical of both the pro-evolution and pro-creation Clergy Letters that are nothing but seeds of division in the world they are required to evangelize.

The problem lies in what the term "evolution" means. Both camps are trying hard to convince the public what evolution truly is when they have no fixed or consistent definition of the term.

It is important to reiterate the following facts. Both evolution and creation are natural processes. At the scientific level, there are microcreation and microevolution. Without microcreation, there can be no microevolution. At the religious level, there are macrocreation, referred to as creationism, and macroevolution, referred to as evolutionism. Both of these fundamentally religious tenets are unobservable in anyone's lifetime. All the successful predictions that scientists claim under the banner of evolutionism are indebted to creationism, and all the failed predictions under evolutionism such as dead genes or "junk" DNA justify creationism. Debates, therefore, between opponents on both sides should be characterized as evolutionism versus creationism instead of evolution versus creationism.

Before delving into an analysis of the Clergy Letters, I want to make sure that the brand of evolution addressed in this discourse is clearly understood from examples of its influence, beginning with Sir Julian Huxley's Darwin Centennial Convocation address on November 26, 1959, at the University of Chicago on the 100th anniversary of *The Origin of Species*. Here is an excerpt from his speech:

> In the evolutionary pattern of thought there is no longer either need or room for the supernatural. The Earth was not created; it evolved. So did all the animals and plants that inhabit it, including our human selves, mind and soul as well as brain and body. So did religion....
>
> Evolutionary man can no longer take refuge from his loneliness in the arms of a divinized father-figure whom he has himself created, nor escape from the responsibility of making decisions by sheltering under the umbrella of Divine Authority, nor absolve himself from the hard task of meeting his present problems and planning his future by relying on the will of an omniscient, but unfortunately inscrutable, Providence.[3]

Huxley's remarks typify the mind-set of a true evolutionist in claiming that evolution is a fact when support of the claim is based entirely on studies at the microevolutionary level. It also is evident from Huxley's address that the term "evolution" carries a religious meaning but presents a scientific front to garner support. James Perloff elaborates on the impact of Huxley's speech as follows:

> That year the National Science Foundation, a U.S. government agency, granted $7 million to the Biological Sciences Curriculum Study (BSCS), which began producing high school biology textbooks with a strong evolutionary slant. Given taxpayer funding, market considerations were no longer a worry. In the 1960s, public schools started using BSCS textbooks. In the meantime, surviving Southern anti-evolution laws were repealed or struck down by the Supreme Court. Students of the sixties thus faced a two-edged sword. On one hand, they were taught evolution, which effectively repudiated God and the biblical version of creation. On the other hand, Supreme Court rulings prohibited teachers from discussing God, reading from the Bible, or praying. It was legal to *deny* God's existence but *illegal* to affirm it.[4]

The topic of evolution in science curricula today clearly threatens and devalues faith in God. The Clergy Letter Project contends that science and religion ask different questions, but is this claim in fact true? The National Academy of Sciences and Institute of Medicine offers this observation:

> Science is not the only way of knowing and understanding.... But science is a way of knowing that differs from other ways in its dependence on empirical evidence and testable explanations. Because biological evolution accounts for events that are also central concerns of religion—including

the origins of biological diversity and especially the origins of humans—evolution has been a contentious idea within society since it was first articulated by Charles Darwin and Alfred Russell Wallace in 1858.[5]

Life's origin is not a scientific problem, but biological diversity is. However, because science and religion now share a similar interest in human origins, can teachers use the science classroom to challenge the religious beliefs of their students? Ex-atheist Gary Parker in *Creation Facts of Life* (1980) shares his fascination with evolutionism and how he displayed it before his students:

> For me, "evolution" was much more than just a scientific theory. It was a total world-and-life view, an alternate religion, a substitute for God. It gave me a feeling of my place in the universe, and a sense of my relationship to others, to society, and to the world of nature that had ultimately given me life. I knew where I came from and where I was going.... I didn't just believe evolution; I embraced it enthusiastically! And I taught it enthusiastically. I considered it one of my major missions as a science professor to help my students rid themselves completely of old, "pre-scientific" superstitions, such as Christianity. In fact, I was almost fired once for teaching evolution so vigorously that I had Christian students crying in my class![6]

Richard Dawkins, unlike many American evolutionists, publicly acknowledges that the Darwinian brand of evolution may lead people and especially children to atheism. In *The Blind Watchmaker* (1987), he narrates his encounter with a Christian student at Oxford University:

I was reminded of the creationist student who, through some accident of the selection procedure, was once admitted to the Zoology Department at Oxford University. He had been educated at a small fundamentalist college in the United States and had emerged a simple, young-Earth creationist. When he arrived in Oxford, he was encouraged to attend a course of lectures on evolution. At the end he came up to the lecturer (who happened to be me), beaming with the primal joy of discovery: "Gee," he exulted, "this evolution! It really makes sense." It certainly does. In the words of a tee-shirt which an anonymous American reader was kind enough to send me: "Evolution—The Greatest Show on Earth— The Only Game in Town!"[7]

Dawkins's use of a science lesson to convert a creationist to evolutionism is evidence that evolution science and religion have similar goals and ask similar questions. Capitalizing on his achievement, one of Dawkins's books is titled *The Greatest Show on Earth: The Evidence for Evolution* (2009).

Why, it might be asked, are evolutionists worried about any form of point-by-point comparison between the creation and evolution worldviews in science classes? It is presumably because such a comparison would likely earn evolutionism the title of myth. Parker shares two of his encounters with evolutionist professors, the first of which concerned radiometric determination of the earth's age.

One of the tensest moments for me came when we started discussing uranium-lead and other radiometric methods for estimating the age of the Earth. I just knew all the creationists' arguments would be shot down and crumbled, but just the opposite happened. In one graduate class, the professor told us we didn't have to memorize the dates of the geologic systems, since they were far too uncertain and conflicting. Then in geophysics we went over all the assumptions that go into radiometric dating. Afterwards, the professor said something like this, "If a fundamentalist ever got hold of this stuff, he would make havoc out of the radiometric dating system. So, keep the faith." That's what he told us, "Keep the faith." If it was a matter of keeping faith, I now had another faith I preferred to keep.[8]

During my graduate studies in geophysics at Ahmadu Bello University, I recall the professor of geology telling the class not to worry about the accuracy of the geological ages. The age of the earth is of no relevance to science and technology except for evolutionists who need it to sustain their worldview. Whether the earth is old or young does not influence science in any way since the laws of science are invariant.

Parker's second experience as a graduate student was an attempt to understand the functioning of DNA, the blueprint of life, in light of the creationism-evolutionism controversy. Parker writes:

All of us can recognize objects that man has created, whether paintings, sculptures, or just a Coke bottle. Because the pattern of relationships in those objects is contrary to relationships that time, chance, and natural physical processes would produce, we know an outside creative agent was involved. I began to see the same thing in a study of living things, especially in the area of my major interest, molecular biology. All living things depend upon a working relationship between inheritable nuclei acid molecules, like DNA, and proteins, the chief structural and functional molecules. To make proteins, living creatures use a sequence of DNA bases to line up a sequence of amino acid R-groups. But the normal reactions between DNA and proteins are the "wrong" ones, and act with time and chance to disrupt living systems. Just as phosphorus, glass, and copper will work together in a television set only if properly arranged by human engineers (as outside creative agent), so DNA and protein will work in productive harmony only if properly ordered by an outside creative agent. I presented the biochemical details of this DNA-protein argument to a group of graduate students and professors, including my professor of molecular biology. At the end of the talk, my professor offered no criticism of the biology or biochemistry I had presented. She just said that she didn't believe it because she didn't believe there was anything out there to create life.[9]

"Belief" is the key word that fits Darwinian evolution, whereas "physical evidence" is the key phrase that characterizes science. Here, Parker explores the field of microcreation which modern scientists have deliberately ignored because it presupposes a creator. The professor's shallow philosophical answer to a brilliant scientific insight by a graduate student is another indication that evolutionism is a religion that is anti-God. Not many professors would want to face intelligent students such as Parker who would expose their ignorance or bias in front of other students: the solution, therefore, is to ban creationist ideas in science classrooms even if this means going to court to enforce this demand.

This begs the question: "With all his doubts about evolutionism, why Parker would upon graduation not remain neutral or become a creationist but rather choose to become an evolutionist in the initial stage of his teaching career?" Here is Parker's response:

Then there is a matter of intellectual pride. Creationists are often looked down upon as ignorant throwbacks to the nineteenth century, or worse, and I began to think of all the academic honours I had, and, to tell the truth, I didn't want to face that academic ridicule.[10]

What could be the faith of a student who refuses to succumb to the indoctrination process? In *Origins: Linking Science and Scripture* zoologist and creationist Ariel A. Roth recounts his graduate-school experience:

When I was a graduate student, the professor of evolution informed me that the faculty of the Department of Zoology was concerned about my creationistic beliefs. He wondered if I could explain them. I responded that I could see how a certain line of thought could lead to a belief in evolution, but that I had several questions about the theory. He was interested. One of the arguments I presented was that I could not understand how the turtle could have evolved from some other reptile without leaving fossil intermediates. The turtle is a unique organism, and in evolving this uniqueness— especially a shell—many intermediates would be involved, yet there is no such evidence in the fossil record. Paleontologists have found thousands of fossil turtles, some almost four meters long. They supposedly evolved some more than 200 million years ago, and in layers below where they first appear, we see no gradual sequence of the evolution of their peculiar shell. After discussing some other considerations, the professor seemed satisfied with my answers and agreed that evolutionary theory had some problems. Later I was told that the only reason the faculty allowed me to graduate was that they could not agree on what to do with me![11]

Because evolutionism is some scientific religion not only students could face the prospect of not graduating as scientists but also staff could be denied tenure, career benefits, or Nobel Prize.[12]

The evolution paradigm, just like any other world religion, has its own version of fundamentalist orthodoxy which pro-evolution clergy are recommending for acceptance as scientific truth. This brand of evolution uses science as a camouflage to impose religious doctrines while suppressing the creation worldview it seeks to replace. In the process, it ostracizes and ridicules other accomplished scientists who differ on philosophical grounds. Its goal ultimately is to promote atheism. Those who are shortsighted see no conflict between evolutionism and inerrant truth. Indeed, Christianity itself is being marginalized because a significant number of Christian clergy in the United States advocate the acceptance of evolutionism as scientific fact and cast suspicion on creationism as ancient myth.

In summary, the brand of evolution that is addressed in this book is the version that Philosopher Michael Ruse presents as an explicit substitute for Christianity and Jesuit priest Pierre Teilhard de Chardin described as a movement that transcends the natural sciences, invading and conquering chemistry, physics, mathematics, and religion.

Problems with the Pro-Evolution Clergy Letter Project

The official objectives and background of the Clergy Letter Project are summarized below:

The Clergy Letter Project is a project that maintains statements in support of the teaching of evolution and in opposition to the teaching of creationism in public schools and collects signatures in support of letters from American Christian, Jewish, Unitarian Universalist, and Buddhist clergy. The letters make reference to points raised by intelligent-design proponents.... This effort was initiated in 2004 by biologist Michael Zimmerman.... The letter was written by the Rev. John McFadden, pastor of the First Congregational United Church of Christ in Appleton, Wisconsin. The Project also encourages congregations to participate in Evolution Weekend by sponsoring events in which clergy and congregations are encouraged to learn about and discuss evolution. The weekend chosen is the closest Sunday to Charles Darwin's birthday, February 12. Evolution Sunday events first took place in 2006, and the Project renamed it in 2008 to be more inclusive.[13]

Michael Zimmerman's address for the February 7–9, 2014, observance of Evolution Weekend reads as follows:

This year marks the ninth annual Evolution Weekend. While all participating congregations will address the relationship between religion and science, many will focus their attention on the theme selected for this year: Different Ways of Knowing/ Asking Different Questions.

Evolution Weekend is an opportunity for serious discussion and reflection on the relationship between religion and science. An ongoing goal has been to elevate the quality of the discussion on this critical topic, and to show that religion and science are not adversaries. Rather, they look at the natural world from quite different perspectives and ask, and answer, different questions.

Religious people from many diverse faith traditions and locations around the world understand that evolution is quite simply sound science; and for them, it does not in any way threaten, demean, or diminish their faith in God. In fact, for many, the wonders of science often enhance and deepen their awe and gratitude toward God.

The importance of the relationship between religion and science is not just a theoretical one. When the science of evolution is mischaracterized for partisan gain, there are very real consequences for society. The theory of evolution, for example, has led to innumerable medical advances, is responsible for amazing agricultural innovations that have helped us feed the hungry, and can provide the knowledge needed to preserve endangered ecosystems. Similarly, given the centrality of evolution to the scientific enterprise, when the theory of evolution is summarily dismissed, the very nature of science is called into question and our educational system is undermined in a dangerous manner.

Similarly, when some define religion so narrowly that it is categorically opposed to evolutionary ideas, or any of the findings of science, it both demeans and diminishes religion. As members of the Clergy Letter Project have stated so often and so clearly, this narrow perspective is at odds with the broader conception of religion held by thousands upon thousands of religious leaders.

Because religion and science use different methodologies to understand the world, and because religion and science ask very different questions, there is no reason to view them in conflict. One important facet of Evolution Weekend 2014, therefore, is to explore the questions each asks and to examine the different ways of knowing embodied in each.[14]

The merit of Zimmerman's undertaking is that unlike his atheistic colleagues such as Richard Dawkins and P. Z. Myers, to name a few, he brings the liberal-minded from both camps into dialogue. To show his disdain for Evolution Sunday celebrations, Myers wrote:

> Today is Evolution Sunday. It's that day when participating ministers will say a few supportive words about evolution from their pulpits, or as I prefer to think of it, when a few people whose training and day-to-day practice are antithetical to science will attempt to legitimize their invalid beliefs and expand their pretence to intellectual authority by co-opting a few slogans. As you might guess, I'm not exactly against the event, but I definitely do not support it. I'm sure a few readers are going to complain that I should be praising these efforts to get people to take baby steps in the right direction, but I just can't do it…. I'm sure the participants in Evolution Sunday mean well and are sincere in their wish to reconcile faith with science, but we'll do far more to promote reason in this country if we withdraw from all participation in the church and let religion wither away from disuse, than we will by encouraging these modern day witch-doctors to spread their delusions.[15]

Given Myers' withering dismissal of Evolution Sunday, Zimmerman scores a point as a better salesperson. However, with his commitment to the National Center for Science Education that honoured him with its Friend of Darwin Award,[16] Zimmerman like his colleagues draws the line between evolutionism and creationism, elevating the former as science and denigrating the latter as religion. Zimmerman's objectives may be summarized as follows: Clergy must accept that evolutionism, which he defines as the theory of evolution, is the foundational scientific truth that underpins modern advances in medicine and agriculture. Opposition to the theory of evolution such as intelligent design thus must not be allowed to conflict with the doctrines of evolutionism. Clergy must rebuke such opposition because to do otherwise is to embrace scientific ignorance and jeopardize religious integrity. In Zimmerman's way of mending the relationship between science and religion, advocates of intelligent design are perceived as the enemies of both science and mainstream religion. With the Clergy Letter Project, then, Zimmerman tactfully shifts the battleground from the science establishment to the religious community or from the laboratory to the altar.

Clerics who have endorsed the project have focused only on the merits of evolutionary theory and not on its shortcomings. By concealing the weaknesses of the theory of evolution the expectation is that with time the public will eventually endorse evolutionism as the absolute truth and view creationism as a myth or a noncomplementary truth. Some of these weaknesses have been discussed openly by evolutionists such as Pierre-Paul Grassé, a renowned French scientist and past president of the Academie des Sciences. Below, for example, are some excerpts from his book *Evolution of Living Organisms* (1977):

> Anyone who endorses the random theory of evolution admits that the eye and the ear, to become what they are, have required thousands and thousands of lucky chances, synchronized with the needs of their construction. What probability is there of such wonderfully fortuitous success?
>
> Natural selection, if one admits that it is the builder of the living world, can only operate if it possesses the correct building materials needed

for the construction of the organ at the right moment. What is the use of appropriate mutations if they appear too early or too late in the course of phylogenesis (the evolutionary development and diversification of particular features of organisms)?... Moreover, during phylogenetic organogenesis, natural selection must be capable of foresight. Isn't "choosing" its prime function? But the choice cannot take place without predicting the future role of the incipient organ. Without such prescience, the coordination of successive states is incomprehensible. Did Darwin take this into consideration?

Without its predictive powers, selection would not be able to favour an incipient organ which, at the time, had little or no usefulness. What sort of advantage could result from the starting of an eye, when the materials forming it were not yet transparent? Of what use was the development of the dentary and the accompanying regression of the proximal jaw bones in theriodont reptiles, the ancestors of mammals? An answer can always be invented, but all this merely adds another supposition to the mass of previous suppositions.

Directed by all-powerful selection, chance becomes a sort of providence, which, under the cover of atheism, is not named but which is secretly worshipped. We believe that there is no reason for being forced to choose between "either randomness or the supernatural," a choice into which advocates of randomness in biology strive vainly to back their opponents. It is neither randomness nor supernatural power, but laws which govern living beings; to determine these laws is the aim and goal of science, which should here have the final say.

To insist, even with Olympian assurance, that life appeared quite by chance and evolved in this fashion is an unfounded supposition which I believe to be wrong and not in accordance with the facts.[17]

Grassé's view that natural laws are central in science's foundation is consistent with that of Albert Einstein, who famously asserted:

I'm not an atheist, and I don't think I can call myself a pantheist. We are in the position of a little child entering a huge library filled with books in many languages. The child knows someone must have written those books. It does not know how.... That, it seems to me, is the attitude of even the most intelligent human being toward God. We see the universe marvelously arranged and obeying certain laws but only dimly understand these laws. Our limited minds grasp the mysterious force that moves the constellations.[18]

I cannot prove to you that there is no personal God, but if I were to speak of him, I would be a liar. I do not believe in the God of theology.... My God created laws.... His universe is not ruled by wishful thinking but by immutable laws.[19]

It is a myth to believe that random chance can generate natural laws. Creation as a primary process and evolution as a secondary process in nature are possible because they originated from a supernatural intelligence that instituted physical laws. Why, then, are many religious leaders complying with Zimmerman's opposition to scholarly examination of the evolution worldview and rejecting the stance of other scientists such as evolutionist Richard D. Alexander, professor of Zoology at the University of Michigan, who advocates the need to compare the evolution and creation worldviews in science classes for purposes of advancing the depth of students' knowledge? Alexander writes with honesty and integrity in illuminating the need for scientific transparency:

> No teacher should be dismayed at efforts to present creation as an alternative to evolution in biology courses; indeed, at this moment creation is the only alternative to evolution. Not only is this worth mentioning, but also a comparison of the alternatives can be an excellent exercise in logic and reason. Our primary goal as educators should be to teach students to think[,] and such a comparison, particularly because it concerns an issue in which many have special interests or are even emotionally involved, may accomplish that purpose better than most others.... In the sense that creation is an alternative to evolution for any specific question, a case against creation is a case for evolution and vice versa.[20]

Some clerics might contend that Alexander is unrepresentative, but other accomplished scientists have voiced similar concerns. For instance, addressing the "Teaching of Origins in Schools," twenty-seven British scientists and educators led by Andy McIntosh, University of Leeds, wrote to the Secretary of State for Education and Skills, Estelle Morris, pointing out the one-sided approach to this sensitive and crucial subject. The letter, published in the *Times Educational Supplement* of April 26, 2002, in part read:

> The National Curriculum requires that Darwinian evolution is put across as the dominant scientific theory but also requires that pupils are taught "how scientific controversies can result from different ways of interpreting empirical data." Science should

be taught with the critical appraisal of alternative theories. Such debate concerning opposing theories provides rigour in scientific method and contributes to the development of critical thinking by pupils.

We find it most inappropriate that some well-meaning scientists have given the impression that there can only be one scientific view concerning origins. By doing so they are going way beyond the limits of empirical science which has to recognise, at the very least, severe limitations concerning origins. No one has proved experimentally the idea that large variations can emerge from simpler life forms in an unbroken ascendancy to man. A large body of scientific evidence in biology, geology and chemistry, as well as the fundamentals of information theory, strongly suggest that evolution is not the best scientific model to fit the data that we observe.

We ask therefore that, where schools so choose, you ensure an open and honest approach to this subject under the National Curriculum, at the same time ensuring that the necessary criteria are maintained to deliver a rigorous education.[21]

Here, scientists as opposed to religious leaders challenge the validity of the theory of evolution on grounds that are wholly consistent with scientific methodology. Let science be science, free from political, social, and religious interference, as an evidentiary path to the discovery of truth.

Integrity in upholding scientific truth against personal conviction is evident in *What Darwin Got Wrong,* an excellent book written by Jerry Fodor and Massimo Piattelli-Palmarini that evolutionists detest. In their preface, they forthrightly declare:

This is not a book about God; nor about intelligent design; nor about creationism. Neither of us is into any of those.... In fact, we both claim to be outright, card-carrying, signed-up, dyed-in-the-wool, no-holds-barred atheists. We thought we'd best make that clear from the outset, because our main contention in what follows will be that there is something wrong—quite possibly fatally wrong—with the theory of natural selection; and we are aware that, even among those who are not quite sure what it is, allegiance to Darwinism has become a litmus for deciding who does, and who does not, hold a "properly scientific" world view.[22]

Therefore, rather than taking sides in the evolutionism-creationism controversy, it would be more prudent for clergy to recluse themselves rather than expatiate on a subject that is outside their area of expertise. Clerics must understand that in the history of science a vigorous minority opinion is often required to overturn a fallacious theory.

In order for science to explain the origin of species without reference to divine creation, it must explain and demonstrate empirically the origin of life. Science has not achieved and cannot achieve that prerequisite. Antony Flew's conversion from atheism to theism was based on arguments at the natural level as opposed to the supernatural. He presents three aspects of nature that point to God:

The first is the fact that nature obeys laws. The second is the dimension of life, of intelligently organized and purpose-driven beings, which arose from matter. The third is the very existence of nature. But it is not science alone that has guided me. I have also been helped by a renewed study of the classical philosophical arguments.[23]

> I must stress that my discovery of the Divine has
> proceeded on a purely natural level, without any
> reference to supernatural phenomena. It has been
> an exercise in what is traditionally called natural
> theology. It has had no connection with any of the
> revealed religions. Nor do I claim to have had any
> personal experience of God or any experience that
> may be called supernatural or miraculous. In short,
> my discovery of the Divine has been a pilgrimage
> of reason and not of faith.[24]

Clerics' view of God is presumably based on faith *only*, that may be why they are putting a boundary between God and science. Would they not be perceived as betrayers if the very evidence from which they are distancing God is what people outside their faith present as a reason for God's existence? For some, particularly students today, the Clergy Letter Project is sufficient grounds for questioning God's existence. Flew, as a professional philosopher, was prepared to follow the truth when the requisite scientific evidence led. Zimmerman, in contrast, as a modern scientist who chose preference over reason is not prepared to follow the evidence and abandon his philosophical conviction; instead, he opts to accept both evolutionism and creationism as different categories of truth.

Zimmerman's view of religion and evolutionary science is controversial. The two disciplines are different in their approach but ask and answer similar questions. What constrains the two is truth.

Therefore, if their conclusions are radically different, then either both of them are false or one of them is false and the other true, but both of them cannot be true as the pro-evolution Clergy Letter Project suggests. The only way we can know which one is false is by comparative study. Based on Parker's experience when he presented an argument against evolutionism and his professor could not challenge it scientifically, one would anticipate that the theory of evolution will crumble if studied with an open mind alongside creationism.

Zimmerman's specious solution, however, is to avoid any form of comparison by maintaining that religion (creationism) and science (evolutionism) ask different questions and provide different answers. Appearing below are rebuttals to Zimmerman's "baits" for enlisting clerics' support in his crusade to justify Darwinism as foundational scientific truth.

Bait #1: Religion and science are not adversaries.

The fact that religion and science are *not* adversaries is the very reason why their precepts should be allowed for study in any classroom whether it is designated as religion or science. Religious worldviews differ, and those of the scientific establishment are divided on philosophical grounds. All those divergences are legitimate fields for objective and disinterested examination.

The relationship between science and religion can only be realized by those whose goal is to establish truth without philosophical boundaries or bias. Einstein once again is brilliantly insightful in this regard:

> Every one who is seriously engaged in the pursuit of science becomes convinced that the laws of nature manifest the existence of a spirit vastly superior to that of men, and one in the face of which we with our modest powers must feel humble. The pursuit of science leads therefore to a religious feeling of a special kind, which differs essentially from the religiosity of more naïve people.
>
> A person who is religiously enlightened, appears to me to be one who has, to the best of his ability, liberated himself from the fetters of his selfish desires and is preoccupied with thoughts, feelings, and aspirations to which he clings because of their superpersonal value. What is important is the force of this superpersonal content, regardless of

whether any attempt is made to unite this content with a divine Being.

Though religion determines the goal, science, in its broadest sense, shows the means for attaining this goal. However, science can only be created by those who are thoroughly imbued with the aspiration toward truth and understanding. This source of feeling, however, springs from the sphere of religion.... I cannot conceive of a genuine scientist without that profound faith. The situation may be expressed by an image: *science without religion is lame*[;] *religion without science is blind.*[25]

With such an honest zeal for scientific truth, Einstein could not endorse the theory of evolution at any point in his lifetime; he was more concerned in how God created the world.

In diametrical contrast to Einstein's humble acceptance of science's limits and the complementary relationship between science and religion is the hubristic and vocal class of scientists who are bent on segregating science and religion in the hope of eliminating the latter. Apparently, the National Academy of Sciences and other such organizations are flooded with atheistic memberships since the acceptance of Darwinism, as Fodor and Piattelli-Palmarini assert, is now the measure of scientific professionalism. Here are a few representative views of modern scientists who believe that science and religion cannot be linked.

It is not that the methods and institutions of science somehow compel us to accept a material explanation of the phenomenal world, but, on the contrary, that we are forced by our *a priori* adherence to material causes to create an apparatus of investigation and a set of concepts that produce material explanations, no matter how counter-intuitive, no matter how mystifying to the uninitiated. Moreover, that materialism is absolute, for we cannot allow a Divine Foot in the door.[26]

Biologists must constantly keep in mind that what they see was not designed, but rather evolved.[27]

Our theory of evolution has become... one which cannot be refuted by any possible observations. Every conceivable observation can be fitted into it. It is thus "outside empirical science" but not necessarily false. No one can think of ways in which to test it.[28]

These views are consistent with Eugenie Scott's assertion that nonmaterial causes are disallowed. Scientists who reason as such are not in search of truth but wish only to protect their philosophical preferences, and they will always interpret the scientific evidence to justify them.

A good example, as previously discussed, is Jerry Coyne's now discredited hypothesis of dead genes or so-called "junk" DNA. In his popular book *Why Evolution Is True*, he made these sweeping pronouncements:

Atavisms (remnants of ancestral features) and vestigial traits show us that when a trait is no longer used, or becomes reduced, the genes that make it don't instantly disappear from the genome: evolution stops their action by inactivating them, not snipping them out of the DNA. From this we can make a prediction. We expect to find, in the genomes of many species, silenced, or "dead," genes: genes that once were useful but are no longer intact or expressed. In other words, there should be vestigial genes. In contrast, the idea that all species were created from scratch predicts that no such genes would exist, since there would be no common ancestors in which those genes were active.

Thirty years ago we couldn't test this prediction because we had no way to read the DNA code. Now, however, it's quite easy to sequence the complete genome of species, and it's been done for many of them, including humans. This gives us a unique tool to study evolution when we realize that the normal function of a gene is to make a protein–protein whose sequence of amino acids is determined by the sequence of nucleotide bases that make up the DNA.... A gene that doesn't function is called a pseudogene.

And the evolutionary prediction that we'll find pseudogenes has been fulfilled amply. Virtually every species harbors dead genes, many of them still active in its relatives. This implies that those genes were also active in a common ancestor, and were killed off in some descendants but not in others. Out of about thirty thousand genes, for example, humans carry more than two thousand pseudogenes. Our genome—and that of other species—is truly a well populated graveyard of dead genes.

A dead gene in one species that is active in its relatives is evidence for evolution. Only evolution and common ancestry can explain these facts.[29]

All of Coyne's predictions have been proven wrong by recent advances in science. In light of the new evidence, clerics cannot say with confidence that "the theory of evolution is a foundational scientific truth, one that has stood up to rigorous scrutiny and upon which much of human knowledge and achievement rests."[30] Darwin posited numerous transitional stages in organic evolution, but those who respect truth today know that there is no such evidence in the fossil record or in the living world. God uses honest scientists to overturn all false predictions.

Bait #2: Evolution is simply sound science that does not in any way threaten, demean, or diminish faith in God.

Evolution is sound science only at the level of microevolution, as in the breeding enterprise. When pro-creation scientists extrapolate microcreation to explain macrocreation, the modern scientific establishment cries foul; however, when pro-evolution scientists extrapolate from microevolution to macroevolution over millions of years, their dogmatic assertions are often accepted as valid scientific fact without proof. Using microevolutionary modifications to explain macroevolutionary transformations threatens, demeans, and diminishes faith in God.

To free pseudoscience from its religious cast, atheistic scientists are careful to avoid the term *evolutionism* and instead speak of *evolution* or *modern biology*. They also use the term *modern science* in lieu of *pseudoscience*, equating the latter with intelligent design. To illustrate this point I have compiled below some quotations from atheist William B. Provine, one of the few American evolutionists who honestly acknowledge that evolutionism demeans Christian beliefs. Unfortunately, he uses code words that prioritize evolutionism. I therefore have italicized the incorrect terminology and enclosed the correct diction in parentheses.

> *Modern science* (pseudoscience) directly implies that the world is organized strictly in accordance with deterministic principles or chance. There are no purposive principles whatsoever in nature. There are no gods and no designing forces that are rationally detectable. The frequently made assertion that *modern biology* (evolutionism) and the *assumptions* (beliefs) of the Judaeo-Christian tradition are fully compatible is false.

As the creationists claim, belief in *modern evolution* (evolutionism) makes atheists of people. One can have a religious view that is compatible with *evolution* (evolutionism) only if the religious view is indistinguishable from atheism.

My observation is that the great majority of modern evolutionary biologists now are atheists or something very close to that. Yet prominent atheistic or agnostic scientists publicly deny that there is any conflict between science and religion. Rather than simple intellectual dishonesty, this position is pragmatic. In the United States, elected members of Congress all proclaim to be religious; many scientists believe that funding for science might suffer if the atheistic implications of *modern science* (pseudoscience) were widely understood.

Liberal religious leaders and theologians, who also proclaim the compatibility of religion and *evolution* (evolutionism), achieve the unlikely position by two routes. First, they retreat from traditional interpretations of God's presence in the world, some to the extent of becoming effective atheists. Second, they simply refuse to understand *modern evolutionary biology* (evolutionist worldview) and continue to believe that evolution is a purposive process.[31]

The goal of the Clergy Letter Project is to weed out intelligent design or creationism from science classes and endorse evolutionism as "modern science." The pro-evolution letters are proof that truth and spiritual insight is independent of people's belief system. Truth is the reason why atheists Fodor and Piattelli-Palmarini, amid strong evolutionist opposition, boldly pointed out that Darwin was wrong. Spiritual insight is the reason why Antony Flew, after fifty years of championing atheism, defied evolutionists' propaganda and endorsed

God's existence without reference to supernatural causation; and why Einstein, although not a conventional believer, consistently portrayed God as the illimitable Spirit behind the natural laws that evolutionists incorrectly credit to chance. In contrast, it is presumably a lack of scriptural spiritual insight that motivates evangelist Reverend Michael Dowd to embrace evolutionism[32] and posit that the "New Atheists" (Sam Harris, Richard Dawkins, and Christopher Hitchens) are God's prophets whom believers should heed.[33] The Scriptures assert that Satan masquerades as an angel of light; hence, one should expect some of his agents to masquerade as agents of righteousness (2 Corinthians 11:13–15).

The ecumenical Clergy Letter Project influences Christians, Jews, Buddhists, and Muslims to honor Charles Darwin for his role in championing a pseudoscientific myth, which they erroneously accept as a scientific and foundational truth. The problem is that the alleged evolutionist truth contradicts the timeless truth of the Bible. Satan is certainly behind the Project's assertion that evolutionism is compatible with the Christian faith. Satan is also behind the Christian clergy's reasoning that rejection of evolutionism is a rejection of the will of our Creator.[34] There can be only one foundational truth in origin studies. To accept both creationism and evolutionism as foundational truths is to heap confusion in the minds of our children. This point resonates with Ken Ham's analysis: "At the same time, atheists take glee when they see the clergy supporting evolution(ism). They usually see such [C]hurch compromise as a step towards atheism, for they expect that the next generation in the [C]hurch will probably see the inconsistency of the clergy's beliefs, and they will soon give up the Bible altogether."[35] This supports Dawkins' assertion that acceptance of the Darwinian theory of evolution leads to atheism.[36]

The only way in which evolutionism can be compatible with Christianity, and hence creationism, is to reverse the whole thrust of the Scriptures, in which case, for example, truth would be falsity, six literal days of invention the same as millions of years of evolution, love interpreted as survival of the fittest, creation by God replaced with evolutionary chance under the instrumentality of natural selection, "sin before death" equals "death before sin," and salvation supplanted by hopelessness. Is it not suspicious as well that it is only on the subject of human origin, which lies outside the domain of empirical science, that academic freedom of inquiry is forbidden under the aegis of the separation of church and state? What our children deserve as a right is the truth about human origin and not the evolutionist myths proclaimed as facts by a preponderance of modern scientists.

Bait #3: Religion and science ask and answer different questions.

Science is the path to truth through physical evidence; religion is the embodiment of truth through divine revelation. Scientific truths are established by natural laws and mechanisms that do not change over time. Its events and processes are testable, reproducible, and repeatable. Theories in science, on the other hand, are opinions on how things are, based on evidence/facts, but these opinions are subject to change upon the emergence of new evidence/facts. Theories, therefore, unlike natural laws cannot be classified as truths at any point in history. To date there is a theory but no law of evolution; hence, there is no foundational scientific truth based on the theory of evolution.

Generally speaking, science and religion look at different aspects of life. Science asks what things are composed of and how they work. On the other hand, religion asks why things exist. On the subject of origins, religion asks and answers why the universe and life exist and what their meaning is. Science can only ask and answer what the material universe is and how it functions. Science can address natural laws such as gravity and the diversity of life, but not singularities such as the origin of life or species that are outside its purview and could impact our basic worldview, our reason for existence, and our hope for the future.[37]

When scientists venture into the domain of religion, they may come up with different interpretations of the same evidence. Let us examine how, regarding genes, scientists can reach conclusions that may have both scientific and philosophical ramifications. Below are two divergent conclusions by evolutionist Richard Dawkins, professor for the Public Understanding of Science, and Denis Noble, professor emeritus and director of Computational Physiology, both of Oxford University.

> Dawkins: [Genes] swarm in huge colonies, safe inside gigantic lumbering robots, sealed off from the outside world, communicating with it by tortuous indirect routes, manipulating it by remote control. They are in you and me; they created us, body and mind; and their preservation is the ultimate rationale for our existence.[38]

> Noble: [Genes] are trapped in huge colonies, locked inside highly intelligent beings, moulded by the outside world, communicating with it by complex processes, through which, blindly, as if by magic, function emerges. They are in you and me; we are the system that allows their code to be read; and their preservation is totally dependent on the joy that we experience in reproducing ourselves. We are the ultimate rationale for their existence.[39]

How genes function is a scientific question, but why they exist is not. For Dawkins, "They created us, body and mind," but there is no scientific evidence that they did. His answer is not a scientific fact but a philosophical opinion. Noble is also an evolutionist but arrives at an entirely different conclusion. The operative philosophical question is "Are genes the reason we exist, or are we the reason genes exist?" Dawkins' conclusion will be more appealing to a highly secularized audience because it credits our existence to a bundle of genes rather than to God. *A body of knowledge with conflicting conclusions cannot be the foundation of science.*

The reason for the conflict is that scientists are not trained to provide expert opinions on philosophical issues, as Flew asserts in explaining his switch from atheism to theism:

> You might ask how I, a philosopher, could speak to issues treated by scientists. The best way to answer this is with another question. Are we engaging in science or philosophy here? When you study the interaction of two physical bodies, for instance, two subatomic particles, you are engaged in science. When you ask how it is that those subatomic particles—or anything physical—could exist and why, you are engaged in philosophy. When you draw philosophical conclusions from scientific data, then you are thinking as a philosopher.[40]

Flew continues:

> Of course, scientists are just as free to think as philosophers as anyone else. And, of course, not all scientists will agree with my particular interpretation of the facts they generate. But their disagreements will have to stand on their own two philosophical feet. In other words, if they are engaged in philosophical analysis, neither their authority nor their expertise as scientists is of any relevance. This should be easy to see. If they present their views on the economics of science, such as making claims about the number of jobs created by science and technology, they will have to make their case in the court of economic analysis. Likewise, a scientist who speaks as a philosopher will have to furnish a philosophical case. As Albert Einstein himself said, "The man of science is a poor philosopher."[41]

The fact that scientists are poor philosophers is evident in their conclusions, such as attributing the existence of computers to abstract intelligence but rejecting the association of intelligence with the origin of DNA, which is in all respects more sophisticated than the best man-made computer.

Below is a comparison between creationism and evolutionism in terms of their common interests.

Table 4.1: A Comparison of Creationism and Evolutionism

Creationism	Evolutionism
Origin of the Universe	**Origin of the Universe**
God by power, wisdom, and understanding created the natural world by • divine command: generating the natural realm, electromagnetic energy, laws, and the elements in the periodic table; • cosmological design: separating land and water, forming constellations, generating solar and lunar energy, imposing fine-tuned constants and producing the various blueprints for plant and animal life with material bodies (the origin of species). The omniscient, omnipresent, and omnipotent God, speaking on Mount Sinai before a live audience, claimed credit for accomplishing the above tasks in six days and rested on the seventh day (Sabbath) as evidence of completing the supernatural creation of the world. God's speech was not delivered by chance. It is not a myth but an eyewitness event. The audience was given three days' notice in advance to prepare for the divine-human conference (Exodus 19:9–25). This historic meeting is the reason the world adopts a seven-day weekly cycle. The seven-day weekly cycle as a creation mandate is the temporal seal/proof of God's ownership of the physical world.	No witnesses of and no consensus on the origin of the universe. The prevailing theory is "The Big Bang" in which concentrated matter and energy of unknown origin accidentally exploded. Correct knowledge of the origin of the universe will never be known. In order to trigger a Mini Big Bang at its laboratory near Geneva, the European Organization for Nuclear Research devoted ten years (1998–2008) to planning and construction at a cost of approximately nine billion US dollars in a project that engaged 10,000 scientists and engineers from over 100 countries.[42] This is concrete proof that to form the universe through a Big Bang requires awesome intelligence and wise planning. Because God is an illimitable Spirit, as Einstein posited, with illimitable intelligence and resources, it would take God only an instant to generate a Big Bang. The late Arthur Ernest Wilder-Smith, a British organic chemist with three doctorates, asserted that because time, entropy, and programming are related, *a mighty intellectual force needs but a moment to accomplish work for which we might need ages.*[43]

Origin of Life	Origin of Life
Fully formed biological organisms were created as unique species and programmed via DNA to reproduce after their kind, subject to changes that include mutational degradation. Life originated from God who is illimitable in knowledge and power. Our experiential knowledge confirmed by the scientific law of biogenesis is that life originates from pre-existing life and not from lifeless matter. Abrupt gaps in the fossil record confirm that what exists in the world today descended from fully formed and discrete species. There is no evidence of transitional links between basic kinds.	Organisms evolved spontaneously from dead matter and progressed from simple to complex forms by random genetic mutations and natural selection. The source of genetic material and the origin of genetic information are unknown. The hypothesis of spontaneous generation violates the scientific law of biogenesis and common sense. The fossil record does not confirm the alleged transitional links in deposits where various organisms are found together. Every claim for the discovery of missing links in isolated sites is disputable and based on sometimes conflicting expert opinions.
Similarities/Dissimilarities among Organisms	**Similarities/Dissimilarities among Organisms**
God used similar materials and design in creating both the DNA and some homologous structures of many organisms. God endowed the various organisms with different non-material virtues (different minds, expression of love, moral values, knowledge, soul, spirit, etc.). This model profoundly explains why living organisms are both similar and dissimilar in many respects.	Organisms are descended from a common primordial ancestor, which logically means that they can only inherit or display similar attributes. A mindless mechanism cannot generate minds or intelligent beings endowed with spiritual values. This is evidence that organisms did not evolve from a common primordial ancestor. This model profoundly explains why organisms are similar but not why they are dissimilar.

These are the facts that students irrespective of their backgrounds must examine to reach reliable conclusions about these two worldviews.

Bait #4: The theory of evolution has led to innumerable medical and agricultural advances.

Successes in medicine and agriculture are confined to the evolutionary limits that govern the breeding enterprise. Jonathan Wells in *The Politically Incorrect Guide to Darwinism and Intelligent Design* (2006) questions ridiculous claims that Darwinism explains everything:

> But if nothing in biology makes sense except in the light of Darwinian evolution, how did it happen that most major biological disciplines were founded either before Darwin or by scientists who rejected his theory? Why do Darwinists claim that their hypothesis is indispensable for agriculture, when it was Darwin who needed farmers—not farmers who needed Darwin? How do Darwinists get away with claiming credit for Mendelian genetics, when Mendel doubted their theory and they ignored his work for decades? In what way is Darwinian evolution indispensable to medicine, when the modern decline in infectious diseases resulted from public health measures and scientific disciplines that owe nothing to Darwin's theory?[44]

The breeding enterprise was already in progress before Darwin, whose contribution to science involved replacing artificial with natural selection in arguing that all species in existence may have a common ancestor.

Publications by evolutionists list numerous successes in science and related fields they attribute to evolutionary mechanisms. A subordinate process, however, is never independent. Without microcreation, there would be no microevolution. As stated earlier, every success attributed to evolution as a secondary process is indebted instead to creation as a primary process. Crediting the innumerable achievements in different fields to evolution merely confirms Pierre Teilhard de Chardin's misguided assertion that evolutionism whose orbit infinitely transcends the natural sciences has successively invaded and conquered the surrounding territory—chemistry, physics, sociology, and even mathematics and the history of religion.[45] This is pure fantasy. In terms of scientific and technological achievements, the world would be where it is today with or without Darwin's theory of evolution. As discussed in Chapter 3, the modern theory of evolution is wrong and portrayed as the biggest mistake in the history of science.

Bait #5: The science of evolution is mischaracterized for partisan gain.

The very nature of science is called into question whenever scientists deliberately refuse to follow the scientific evidence wherever it leads. When one does not pursue the truth to the very end, he or she does so either for fear of losing something or in the hope of gaining something. Both ulterior motives characterize zealous advocates of evolutionism. Science's inability to show how life can be generated spontaneously is a major obstacle for the evolutionist. Louis Pasteur's findings that life can only derive from pre-existing life are a knockout blow, as Hubert Reeves et al. explain:

> Because of him [Pasteur], scientists concluded that life could not come directly from inert matter; therefore, it could come only from life itself. Which raised the essential question: how do you explain the initial manifestation of life? There were only three solutions: divine intervention, which removed the matter from the hands of science; chance—in other words, some kind of accident—which took the matter into the realm of miracle, which is difficult to accept; or an extraterrestrial origin—germs of

life that were brought here by meteorites—which didn't solve the question either.[46]

Since miracles are not part of science's purview, modern scientists are left with the option of choosing between (1) scientific integrity by following the evidence of divine intervention and (2) scientific preference by advocating evolutionism for partisan gain. Majority of modern scientists incorrectly perceive divine intervention as a loss to the scientific enterprise, but they fail to realize that a superior scientific mind and not chance or selection is behind the abundance of scientific resources and knowledge. The fact that God created light does not stop scientists from studying the nature of light and its properties. For instance, Sir Isaac Newton as a devout Christian believed that God created light but that did not stop him from pursuing his scientific investigations. He was instead fascinated about light and colours and became the first person to understand the true nature of the rainbow and his theory of light laid the foundation for modern physical optics. Divine intervention, therefore, does not remove any matter from the hands of scientists who are after true knowledge and are prepared to follow the scientific evidence wherever it leads without preconditions. In this regard, the late, Nobel laureate Professor George Wald, in 1954, wrote "I think a scientist has no choice but to approach the origin of life through a hypothesis of spontaneous generation"[47]; but in 1967 reached the conclusion "It is mind that has composed a physical universe that breeds life, and so eventually evolves creatures that know and create: science-, art-, and technology-making animals."[48] Wald was an evolutionist who recognized the limits of scientific inquiry and eventually gave up his preference (spontaneous generation) to embrace reality (the view that the physical universe was composed by a mind).

Partisan gain is the reason why Zimmerman does not support the need for students to analyze Darwinism's strengths and weaknesses in science classrooms. Which is the more responsible course to follow: to solicit clergy to promote evolutionism to students in their congregations, which is tantamount to indoctrination; or to allow students in science classrooms to compare divergent worldviews, which is an act of academic training in any discipline? Evolutionists like Zimmerman will avoid the latter because, as the saying goes, they live in glass houses.

Problems with the Pro-Creation Clergy Letter Project

According to Reverend Kyle Ann Lovett, one of the signatories of the pro-creation Clergy Letter Project, "The Bible is a rich and deep and profound collection of texts. But it's not written to be a scientific text, or a recipe book. It's got a wide range of writing styles and approaches: poetry + wisdom writings, cosmic inspiration + a theological interpretation of a piece of human history. So do the other sacred religious texts and the Creation stories that abound in our world."[49] What makes the Bible unique from other Creation stories, however, and to some degree compatible with science is that it is also a compilation of laws that can only be interpreted literally. Those biblical texts that establish divine laws, particularly the moral laws not mentioned by Reverend Lovett, are timeless and parallel scientific laws that are invariant. The Creation Sabbath law (Exodus 20:8–11) that I have discussed earlier is one such binding commandment. The relevance of this law in establishing the world's seven days weekly cycle is the reason why creation was accomplished in six days.

This book's main criticism of the pro-creation Clergy Letter is its failure to acknowledge the oral and written testimony by Jehovah/ God concerning the Creation Sabbath law. The six days of creation in Genesis are presented in the form of a story, while the six days described in Exodus 20:8–11 constitute a moral commandment. On which of the two should an unbiased Christian focus? The answer is: both. Because the Holy Bible communicates truth, it is wise to rely on both since they ensure consistency on the subject of divine Creation. Unfortunately, a majority of Christians today believe that the Creation Sabbath commandment is no longer binding because they associate it with the Jewish observances that Christ abolished by his sacrificial death and resurrection. Their primary focus is on the Genesis account, as if that in Exodus is of no significance. For instance, in two important articles on Genesis the Christian Answers Network ignored God's institution of the Sabbath law. The first article titled "Should Genesis Be Taken Literally?"[50] addressed the Bible's genres of poetry, parables, prophecy, letters, and testimony but made no reference to God's claim before a live audience of having created the world as one of authentic historical evidence. The second article titled "Does God Expect Us to Read Genesis 1–11 as a Record of Authentic Historical Fact?"[51] cited internal evidence from the Book of Genesis and Evidence from the rest of the Bible but did not mention God's personal claim contained in the Decalogue. Such a claim is more authoritative than a story, and without taking it into account, Genesis is subject to endless controversy. Forsaking to observe God's Creation Sabbath day in compliance with church tradition should not be a valid reason to omit or neglect discussing its significance as a creation ordinance.

Although the Genesis account differentiates days from years, many biblical scholars ignore this fact to suggest that the "day" implied in the commandment is not a period of twenty-four hours.

They support their views by referring to 2 Peter 3:8, which asserts "one day is with the Lord as a thousand years, and a thousand years as one day." This verse simply means that on the subject of *patience* our timing is irrelevant to God, not that God does not know the literal difference between one day and a thousand years. God established both a seven-day Sabbath to commemorate initial Creation and a seven-year Sabbath for agricultural purposes. Also, before declaring on Mount Sinai the Creation Sabbath commandment that stipulates six days of creation, God gave the children of Israel three days' advance notice to prepare for the occasion (Exodus 19:10–11). God then showed up on the third day (Exodus 19:16–19) and not after three thousand years. To argue that the six days specified in the Decalogue's moral laws has no literal meaning is to undermine God's integrity. It was to eliminate any element of doubt that God offered to speak to Moses before an audience so that they would believe Moses and every statement in the Decalogue (Exodus 19:9,19).

Any defense of God's six days of creation in Genesis without reference to the Exodus Sabbath law is inadequate for many skeptics. In a recent book, *Creation or Evolution? Origin of Species in Light of Science's Limitations and Historical Records* (2011), I presented the Exodus evidence supporting God's claim of having created the world. The attempt resulted in the following comments from *Kirkus Reviews*: "Ebifegha takes the theory of evolution for a ride and finds it wanting and in need of some creative intelligence. Ebifegha claims that another approach has already answered that question, intelligent design, wherein we find 'the evidence of creation, including God's written and verbal claim for having created the universe.' As an approach to understanding immateriality, that is only one, and a rather limiting, course. Ebifegha is preaching to the converted; his creation/creationist stance leaves no room for debate."[52] A personal claim by God, who is the author of truth, brings the creationism-evolutionism debate to a close for those who seek after the truth.

In a July 2002 article titled "15 Answers to Creationist Nonsense," John Rennie as editor of *Scientific American* challenged creationists by writing, "If superintelligent aliens appeared and claimed credit for creating life on earth (or even particular species), the purely evolutionary explanation would be cast in doubt. But no one has yet produced such evidence."[53] Responses to his implicit challenge have been shallow.[54,55] However, omniscient God provided the answer several thousand years ago. The Creation Sabbath law is timeless and authenticates God's sovereignty over the universe. Unfortunately, again, a majority of Christians wrongly perceive it as a superseded and purely ceremonial Jewish custom. In the Decalogue the weekly Sabbath or the seventh day is formally presented as the *Sabbath of the Lord* (Exodus 20:10). Jesus Christ perceived observance of the weekly Sabbath as a divine mandate on creation and not as a Jewish custom and this explains why he told the Pharisees that *he is the Lord of the Sabbath* (Mark 2:28). Because Christ came to save the Jews and others, both are formally welcomed to keep the Sabbath (Isaiah 56:1-8) as a universal covenant. Christ did not change the order of creation or abolish creation, which means he did not change the day of the Sabbath or abolish the Creation Sabbath day.

The folly of the pro-creation Clergy Letter is that, like the pro-evolution Clergy Letter, it is incomplete without reference to the Creation Sabbath law. God instituted it for a purpose, and in the modern world, it provides a bulwark against any false worldview. Without the Sabbath law on creation as the ultimate authority, the pro-evolution and pro-creation Clergy Letters are basically religious channels that respectively promote the evolutionist and creationist arguments proceeding from the scientific establishment. In the next chapter we will examine a typical sermon delivered on Evolution Sunday.

5

Analysis of A Pro-Evolution Sermon

These people draw near to Me with their mouth,
And honour Me with *their* lips, But their heart is far
from Me. And in vain they worship Me, Teaching
as doctrines the commandments of men.

—Matthew 15: 8-9 (NKJV)

The pro-evolution clergy resolution, *"We ask that
science remain science and that religion remain religion,
two very different but complementary form of truth,"*[1]
prohibits the celebration of Darwin's theory of
evolution on Church premises.

Lessons on Darwinism advocating
evolutionary bacteria-to-human transformations
in science classrooms marginalize the essence of
God's role in creation; promoting the theory of
evolution as foundational truth in Church buildings
questions the rationale behind any redemption
formula. For most youths in the modern world the
God-free Atheist's Sunday Assembly whose motto

is "Live Better, Help Often, Wonder More,"[2] seems to make more sense.

—Michael Ebifegha

The Evolution Weekend sermons of 2014 are unique because they address "Different Ways of Knowing/Asking Different Questions." I have chosen to examine only one sermon because of space limitations and because it is a good representation of the pro-evolution cleric mind-set.

Below is an extended excerpt from a sermon by Dr. Jill Joseph, Senior Warden of All Saints Episcopal Church in Sacramento, California, that was delivered on February 9, 2014 in honor of Charles Darwin's birthday.

> Today we celebrate Evolution Sunday with the theme this year of "A Different Way of Knowing."…
> I begin these comments with two cautions.
>
> First, as I understand our faith and creed and sacraments, we both hold them dear and must always acknowledge our own ignorance and the unknowability of God…. The second caution is this: our science is, almost by definition, incomplete, and can also be terribly misused. As you can probably safely assume by now, I will argue that evolution is real and, more generally, that science is a gift.
>
> But scientific knowledge is incomplete, and the debate continues regarding many issues of fundamental importance. What events created the cataclysmic beginning we call the "Big Bang"? How did life on Earth first arise? Is there life elsewhere in the universe?
>
> Having shared these cautions, let's begin by considering the foolish and false choices that

are created when we hold an image of God that exists only where rational thought and scientific explanation fail.... If mountains trembled, it was the message of God's power directed to us. No further search for explanation was needed.... This creates, of course, an irreconcilable but entirely false choice between science and faith.

Ironically, our fundamentalist fellow Christians and our atheist neighbours agree on this choice between science and faith, although they come to very different conclusions.

The atheist sees that scientific knowledge can describe so much about our origins and places us as one amid many creatures and concludes that therefore God cannot be real. There is no place for God because there is nothing missing, nothing left to explain, no place where rational thought and scientific explanation fail.

The fundamentalist agrees that either God or science must prevail, and therefore concludes that science cannot be real....

Fortunately, we gather and worship in the Episcopal tradition and hold dear the words of Jesus who described the greatest commandment as that of loving God with all your heart and with all your soul... and with all your mind. For myself, I take this to mean we are not only invited, but indeed obligated, to incorporate rational inquiry and intelligent discernment provided by modern science into our faith. Fortunately, we are not trapped by the false dichotomy between faith and science.

We are capable of understanding that the wonderful story of creation that begins the Hebrew

Bible with the words, "B'rishit... in the beginning God created the heavens and the earth," is actually two quite contradictory stories that emerge in the first and second chapters of Genesis.

We are not troubled by this inconsistency.

We do not need to debate whether God did or did not take six days or one day for the creation of the Earth.

We need not be perturbed by pondering whether all living things, plant and animal, were either created before woman or after.

This does not concern us because we know that in Genesis we are not seeking the language of scientific fact, but rather participating in a different way of knowing....

Consider this:

We can observe that every galaxy in the universe is being flung away at immense speeds from an incomprehensibly gigantic blast that we refer to as the Big Bang.

13.8 billion years ago this Big Bang exploded with inconceivable energy, and from a mass the size of a pebble flung forth all matter and energy and space and time that now exist as 500 billion galaxies in the known universe....

In the first billion years or so, meteors and comets bombarded the Earth, melting their icy cores, blasting open geothermal vents, creating places where water and energy mixed with the chemical building blocks that could become amino acids and life itself....

And primates emerged, small and scurrying at first, then slowly adapting to changing climate and growing larger, smarter.

One of those was our Neanderthal cousins, and we know now that around 2% of our genetic code is derived from those pre-humans relatives.... Through evolution we have a wide-ranging relationship to all of life on earth....

Every living cell of every organism goes about its daily work of creating and copying genetic information, of shuttling waste, of creating enzymes, of building and repairing its own structures, of creating and transmitting energy. This continuity is the gift of evolution that has selected for efficiency....

Evolution and science are, as suggested by the title of this homily, simply another way of knowing that does not compete with my faith. But they are, for me, a precious way of knowing.

The questions that I ask in faith are other questions. The realities I touch in faith are other realities. I don't turn to God to find out how the universe was created or how old the Earth is.

I turn to God to be tutored in love, to be held accountable for the state of my soul, to receive the grace of forgiveness, to learn how to live and then how to die.

Science relies on honest inquiry. Faith relies on honest prayer. They ask different questions; they provide different answers. They are different ways of knowing....[3]

By way of analyzing this lofty sermon, we must first establish some crucial points. First, evolution is a natural process. It is neither ancient nor modern science. Evolutionism, on the other hand, is pseudoscience because like creationism, it is a belief that cannot be empirically confirmed, and as a contemporary movement, its objective is to sidetrack rather than complement scientific inquiry. Its advocates believe that it is a movement that transcends empirical science and hence free to violate scientific laws. While science relies on honest investigation without preconditions, evolutionism comes with the precondition that materialism is absolute and that nothing else such as spirituality or immateriality counts. Given this precondition DNA, life's blueprint, is a random product of chance and required no intelligent designer for its emergence. Such a conclusion is nonsensical, dishonest and unscientific.

Second, Dr. Joseph asserts in the final paragraph quoted above that "Faith relies on honest prayer." One is curious to conceive of what constitutes a dishonest prayer and what the rationale for such a prayer might be. For instance, one might ask, "Is Christ's valedictory prayer offered before his disciples in John 17:1–5 an honest or dishonest prayer?" It reads as follows:

> Father, the time has come. Glorify your Son, that your Son may glorify you. For you granted him authority over all people that he might give eternal life to all those you have given him. Now this is eternal life: that they may know you, the only true God, and Jesus Christ, whom you have sent. I have brought you glory on earth by completing the work you gave me to do. And now, Father, glorify me in your presence with the glory I had with you before the world began.

Prayers may have different motives, but each is honest. In light of Christ's valedictory prayer above, why do clergy need an Evolution Sunday to address our origins? Faith is based on trust and conviction. The Webster's Universal Dictionary & Thesaurus (Geddes $ Grosset, Concord, Ontario, 2003) defines faith as "trust or confidence in a person or thing; a strong conviction, esp. a belief in a religion; any system of religious belief; fidelity to one's promise, sincerity."[4] Lack of trust or faith in God is the reason why one would not turn to God for answers about an unobservable event. Far too many pro-evolution clergy credit the fallacious assertions of atheistic scientists.

Third, we must be careful not to conflate faith with superstition. Whereas faith is trust based on religious knowledge or scientific evidence, superstition is uncertainty/fear based on the absence or lack of any religious canon or scientific insight. Statements, such as "If mountains trembled, it was the message of God's power directed to us" are under the genre of superstition and, hence, are irrelevant to the evolutionism-creationism controversy.

Fourth, love for God is not determined by prayers or communal worship services but by obedience to God's will. God reminded King Saul that obedience is preferred over worship (1 Samuel 15:22). In proclaiming the Ten Commandments God maintained that obedience in keeping them is the evidence of love (Exodus 20:6). In presenting six days of creation as one of the commandments in the Decalogue, God made acceptance of this fact a moral obligation.

Christ taught that the first and greatest commandment is to "Love the Lord your God with all your heart and with all your soul and with all your mind." And the second is like it: "Love your neighbour as yourself" (Matthew 22:37–38 NIV).

Accepting Jesus as a personal saviour and loving Jesus are two different things. 1 John 5:3 stipulates, "For this is the love of God, that we keep his commandments," and in John 14:15 Jesus said to his followers, "If you love me, keep my commandments." God's commandments are literal. The Creation Sabbath commandment in the Decalogue addresses our relationship with God and neighbours. Obedience in honouring God's Creation Sabbath law therefore demonstrates (1) trust in God's claim of having created the world in six days followed by a seventh day of rest, and (2) love of God as the Creator and, correspondingly, love of others as fellow creatures. Believers who keep the Sabbath as God's Creation ordinance, as opposed to a Jewish custom, have *divine* immunity from any competing worldview, whereas those who do not are vulnerable to false worldviews. To reject God's Sabbath commandment as a myth does not reflect love of the Creator. God's directive to keep His commandments ensured that the truth about our origins was not exclusively in the hands of scientists or theologians. Having addressed these concerns, we can now turn to a closer examination of Senior Warden Joseph's message on Evolution Sunday.

What is at stake here is not what Joseph believes but how her belief will impact the gospel of Jesus Christ. The gospel's foundation is creation, with Jesus Christ as the cornerstone. The new foundation that Joseph and pro-evolution clergy are laying is based on evolution, with the new atheist scientists as the cornerstone. Are such clerics saying that God cannot distinguish between creation and evolution as processes in nature? It comes down to two things: either (1) God and Jesus, who both spoke about a literal Adam, are incapable of differentiating between the processes of creation and evolution, or (2) God purposely misled human beings on the subject of origins in claiming through a moral law to have created the world in six days.

The positive aspect of Joseph's sermon is that, unlike the Clergy Letter Project's manifesto, she mentions some of the difficulties science has not solved and is perhaps incapable of solving. This positive dimension, however, is qualified by her lengthy presentation of evolutionist tenets. Concerning creationism, she points out only what she deems as inconsistent between Genesis 1 and 2. She also fails to inform her audience of the fact that numerous evolutionary speculations are conflicting and subject to revision upon the emergence of new scientific data. This does not come as a surprise because once cleric is not troubled by religious inconsistencies he or she is bound not to be troubled by any degree of scientific inconsistencies. For Joseph, creationism has no connection to science, which of course is complete nonsense. The raw material for evolution as a secondary process must first have been created. It therefore would be more prudent for her to posit that every organism's "daily work of creating and copying genetic information, of shuttling waste, of creating enzymes, of building and repairing its own structures, of creating and transmitting energy" is the legacy of *creation* as opposed to *evolution*.

Because evolution is a consequence of creation and not vice versa, one might ask whether Joseph's claims are consistent with the will of God who does not share His glory with unfounded myths, as we read in Romans 1:18–32 (NIV):

> The wrath of God is being revealed from heaven against all the godlessness and wickedness of men who suppress the truth by their wickedness, since what may be known about God is plain to them, because God has made it plain to them. For since the creation of the world God's invisible qualities— his eternal power and divine nature—have been clearly seen, being understood from what has been made, so that men are without excuse. For although they knew God, they neither glorified him as God nor gave thanks to him, but their thinking became futile and their foolish hearts were darkened. Although they claimed to be wise, they became fools and exchanged... the truth of God for a lie, and worshipped and serve created things rather than the Creator.

Joseph's otherwise captivating presentation of evolutionism to congregants who respects her views and her credentials as an MD and senior warden reminds one of scholar Pierre-Paul Grassé's observation:

> Today our duty is to destroy the myth of evolution, considered as a simple, understood, and explained phenomenon which keeps rapidly unfolding before us. Biologists must be encouraged to think about the weaknesses and extrapolations that theoreticians put forward or lay down as established truths. The deceit is sometimes unconscious, but not always, since some people, owing to their sectarianism, purposely

overlook reality and refuse to acknowledge the inadequacies and falsity of their beliefs.[5]

It also is noteworthy that Joseph's sermon violates the Clergy Letter Project's recommendation: "We ask that science remain science and that religion remain religion, two very different, but complementary, forms of truth." Indeed, given its few references to God and Jesus, Dr. Joseph's sermon could be construed as a lecture on evolutionism in an advanced science class.

Interestingly Joseph asserts, "We do not need to debate whether God did or did not take six days or one day for the creation of the Earth." This debate, of course, is not necessary because God proclaimed the six days of Creation to a live audience and in print on stone tablets. We thus do not have to debate whether divine Creation is a foundational scientific truth since God is the greatest scientist.

In order to convey her evolutionist convictions to her audience, Joseph first had to suggest that the Genesis 1 and 2 accounts are contradictory. This is the normal approach by theistic evolutionists. They deliberately ignore the Exodus account where the six days of Creation serve as the platform for an unequivocal commandment. Dr. Ralph Walker, Emeritus Fellow and Lecturer in Philosophy at Magdalen College Chapel, Oxford, follows the same line of argument as Dr. Jill Joseph in his Evolution Sunday sermon on February 9, 2014:

> Before Darwin, a great many Christians, Jews and Muslims took literally the biblical account of God's creation of man and woman quite separately from his creation of animals. Scholars had of course long been aware that the Bible offers two contradictory accounts of this creation, and that to take it literally is to misconstrue its proper purpose; but what was familiar to scholars had not percolated through to the ordinary coalminer or bishop. Darwin's theory of evolution showed that we are the products of

a long evolutionary process. This by itself caused
those who would listen to rethink what the real
claims of their religion are, getting rid of the many
spurious claims that had become attached to it over
the centuries, and returning to its core message
about God's relationship to us and ours to Him.[6]

First, Darwin's theory of evolution has not shown us that we
are the products of a long evolutionary process; it is a belief that has
not and cannot be proven. Second, God's relationship to us is based
on covenants, signs, and laws. The Creation Sabbath is an everlasting
covenant, a sign of holiness and, hence, of affiliation with the God
of Abraham, Isaac and Israel, and evidently a moral law by divine
classification (Exodus 20:8–11, 31:12–18). Evolution is not one of the
terms of God's covenant. Third, Genesis 1 and 2, only appear to be
contradictory when clerics and scholars misconstrue both accounts.

While Genesis 1 presents the "macro" events during each of the
six days of Creation, Genesis 2 is a more detailed explanation centred
on humankind and its primal environment. For instance, Genesis 1
stipulates that man and woman were created on the sixth day, but
Genesis 2 specifies that man was created first and then woman. Genesis
1 also stipulates that the seas were divided from dry land on the third
day, while Genesis 2 reveals the names of four historic rivers flowing
outwards from Eden—Pishon, Gihon, Tigris, and Euphrates—with no
reference to the day(s) on which they were formed. Genesis 2 conveys
the following information[7]:

- Adam is created (2:7)
- Garden of Eden created (2:8–9)
- Description of river system in Eden (2:10–14)
- Adam put in Garden and given instructions (2:15–17)
- Adam names some kinds of animals (2:18–20)
- God creates Eve (2:21–22)
- Description of Adam and Eve's marriage (2:23–25)

Pro-evolutionists are quick to compare Genesis 1 and 2 but reluctant to point out the consistency in the number of days of Creation between Genesis 1 and Exodus 20:8–11. Genesis 2 does not contradict Genesis 1 except in the minds of skeptics who choose to wonder whether plants and animals were created before or after woman. Some skeptics contend that Genesis 2 indicates that the creation of everything was accomplished in one day. God counters this hypothesis in the Creation Sabbath commandment according to which the six days of primal creation is reaffirmed. Pro-evolution clergy, consequently, use Genesis 2 simply to support their argument that the Bible cannot be taken literally.

When a conflict arises between the Bible and science on an event that no human being witnessed, the former should prevail. For instance, Walkers' assertion that "Before Darwin a great many Christians, Jews, and Muslims took literally the biblical account of God's creation of man and woman quite separately from his creation of animals,"[8] undermines the truth in Genesis that stipulates the following:

> Then God said, "Let Us make man in Our image, according to Our likeness; let them have dominion over the fish of the sea, over the birds of the air, and over the cattle, over all the earth and over every creeping thing that creeps on the earth." So God created man in His own image; in the image of God He created him; male and female He created them (Genesis 1:26-27 NKJV)

Was God here invoking evolution by natural selection such that human beings would emerge via numerous transitions? Or this verse is consistent with Christ's valedictory prayer. Christ came to save those created in the image of God. Unless Christ also is the product of evolution, we need no grace of forgiveness if we indeed evolved as some modern clergy now believe. We need not turn to God in order to learn how to die, as Joseph suggests, because death is no longer the consequence of sin.

Joseph's assertion that "I don't turn to God to find out how the universe was created or how old the earth is" is also unscriptural and rather naive. God is the ultimate authority on events beyond human ken such as the origin of life. Those who claim that the earth is thousands to billions years old are wrong to claim that their data is based on the biblical record. In particular, the claim by Jehovah Witnesses that "The Bible reveals that the creative 'days' or ages encompass thousands of years" is wrong.[9] All we know is that the planet was summoned into existence. God does not tell us how old it is and science cannot tell us its structural age because it falls outside its purview.

It is in response to false teachings that infiltrated the ancient church and that now are besieging the modern church that the Apostle Paul wrote:

> I am astonished that you are so quickly deserting the one who called you by the grace of Christ and are turning to a different gospel, which is really no gospel at all. Evidently some people are throwing you into confusion and trying to pervert the gospel of Christ. But even if we or an angel from heaven should preach a gospel other than the one we preached to you, let him be eternally condemned! As we have already said, so now I am saying again: If anybody is preaching to you a gospel other than what you accepted, let him be eternally condemned! Am I now trying to win the approval of men or of God?… Or am I trying to please men? If I were still trying to please men, I would not be a servant of Christ. (Galatians 1:6– 10)

It is the prayer of everyone who loves God with all his or her heart, soul, mind, and strength that modern liberal clergy, in an effort to please scientists and be politically correct, should not mislead the public they are ordained to evangelize. Clergy have only two choices: accept God's personal claim of having created the world as truth or reject it as myth. There is no middle ground. There is only one truth, not two conflicting truths, on the subject of our origins. In the next chapter we will examine the consequences of the pro-evolution Clergy Letter.

6

Consequences of The Pro-Evolution Clergy Letter Project

Contrary to clergy's opinion, our knowledge of the origin of species can only be valid and complete when both science and religion profess the same truth, as opposed to two radically different truths. Either human beings are the product of millions of years of evolutionary transformations from bacteria of unknown origin, or they were created uniquely by God as revealed in Genesis 1:26-27, claimed in Isaiah 45:12, and legislated in Exodus 20:8-11; 31:14-18. The current clerical proposition that both are different truths is false and not the honest solution to the evolutionism-creationism controversy.

In promoting evolutionism as scientific truth, clergy are relegating basic religious truths, including God's moral commandment on Creation in the Decalogue, to the status of myths and elevating instead macroevolutionary theory as truth. Science is, neither theistic nor atheistic. The everyday use of science fulfills these conditions and therefore belongs to the State and its inhabitants. Origin science, on the other hand, involves either theistic assumptions (such as life arising from pre-existing life, which is consistent with the natural law of biogenesis that points to God) or atheistic assumptions (such as life arising from non-life, which is contrary to natural law and inconsistent with experiential knowledge). Both worldviews are religious positions. In the modern world there is no longer "separation of church and state" but instead "separation of worldviews from state policies." This suggests that no worldview, neither creationism nor evolutionism, should be taught in public schools if the State wants to forbid any form of indoctrination.

—Michael Ebifegha

God created the world in six days and included the Sabbath as a divine seal. This seal comprising the seventh day block of holy time is immaterial, incorruptible, universal, and permanent. This means God does not have to renew or present the claim for having created the world before another audience just in the same manner that Christ's death on the cross for the salvation of the world is sufficient for all generations.

The reason for Christ as God's gift is to link salvation to God; the reason for the Sabbath Day as God's gift (Ezekiel 20:12) is to link Creation to God. While to accept salvation is a choice, to be born is not. Hence, while salvation is free, observance of the Creation Sabbath law is incumbent on everyone as a moral law. Creation and redemption are two sides of the same coin. If the story of Creation is a myth as some clergy claim, it follows that the redemption story is another myth, and the gospel simultaneously loses both its foundation and power.

The following questions that scientist Francis S. Collins poses in *The Language of God* (2006) are noteworthy: "Is not God the author of the laws of the universe? Is He not the greatest scientist? The greatest physicist? The greatest biologist?"[1] If the answer to these questions is yes, the most intriguing question that follows is: Why, then, did God not mention evolution if it is a foundational truth? If evolution is foundational, what is the rationale for Christ's redemptive mission? If the Creation Sabbath commandment is a myth, why did Christ foster its observance (Luke 4:16)? These concerns suggest that in endorsing evolutionism as a foundational truth liberal clergy are setting a new foundation independent of the Scriptures they inherited from the prophets, the apostles, and Jesus Christ as the chief cornerstone (Ephesians 2:19–20), all of whom affirmed the six days of creation as literal.

The primary reasons for endorsing the Clergy Letter Project are discussed by Albert C. Kuelling, a retired physicist who wrote two of the petitions related to science and faith that were accepted at the United Methodist Church's 2008 General Conference. He wrote:

> From the religious mistakes of the late Middle Ages, many denominations have learned that repressing science is often counterproductive to good religion. In 1600, the Church burned Italian cosmologist Giordano Bruno at the stake for speculating that intelligent life might exist somewhere beyond Earth.
>
> Shortly after, Galileo was threatened with the same fate during his inquisition. His heresy? Teaching that the Earth is not the center of the universe. Galileo recanted to save his life, but he was put under house arrest for the last decade of his life.
>
> Learning from the past, some denominations now understand that new findings about the natural universe often represent new revelations of the mysteries of God's Creation and Word. They

increase, rather than detract from, our awe and reverence for almighty God's capabilities.

At the 2008 General Conference, three petitions made the following changes to United Methodist documents:

1. Petition 80050: accepts evolution and corrects some ambiguities under "Science and Technology" in the *Book of Discipline.*
2. Petition 80990: endorses the Clergy Letter Project and its reconciliatory programs between religion and science and urges United Methodist clergy participation, in Resolution 11, "God's Creation and the Church," in the *Book of Resolutions.*
3. Petition 80839: creates a new resolution, "Evolution and Intelligent Design," in the *Book of Resolution*: "The United Methodist Church goes on record as opposing the introduction of any faith-based theories such as Creationism or Intelligent Design into the science curriculum of our public schools."

Historically Methodism has sidestepped honest dialogue about the interface between religion and science, especially about evolution. This appears to have been done out of fear that accepting the findings of science— for instance, that evolution is an established scientific cornerstone, especially in biological fields—might incur the wrath of creationists within Methodism.

The resulting effect had been an implication that the National Academy of Science and hundreds of scientists worldwide over the last century and a half are wrong.

The large voting percentage in passing the three evolution petitions is evidence that the leadership of the United Methodist Church recognizes the need to change this situation. So Methodism is joining many other denominations around the world that find no conflict between religion and science.

Many young folk have left the church because they have not been thoroughly grounded in the understanding that God uses metaphors, beautiful stories, and other means to enhance understanding of religious principles. Thus, when there appears to be a conflict—albeit a needless one—between religion and science, they quietly leave.

Young people typically don't say how important this issue is to them because they don't want to insult those they leave behind by saying that their religion is out of touch with reality.[2]

First and foremost, the phrase "faith-based theories such as Creationism or Intelligent Design" needs to be addressed. *Microcreation* and *microevolution* are scientific theories, but *macrocreation* (special creation/intelligent design) and *macroevolution* (molecule-to-human being transformations/unintelligent design) are faith-based theories. Therefore, in advocating the teaching of evolutionism in science classes and banning the teaching of creationism or intelligent design, clergy are expressing their faith in evolutionists. In believing that "the National Academy of Science and hundreds of scientists worldwide over the last century and a half" cannot be wrong, clerics are affirming their faith in evolutionism; in contrast, they are contending that God must be wrong in claiming that Adam was created on the sixth day (Genesis 1:26–31; 2:7). Their absolute trust in scientists does not come as a surprise. The National Academy of Sciences and Institute of Medicine, in its 2008 booklet titled *Science, Evolution, and Creationism*, made the following declaration:

> Many scientific theories are so well established that no new evidence is likely to alter them substantially. For example, no new evidence will demonstrate that the Earth does not orbit around the Sun (heliocentric theory), or that living things are not made of cells (cell theory), that matter is not composed of atoms, or that the surface of the Earth is not divided into solid plates that have moved over geological timescale (the theory of plate tectonics). Like these other foundational scientific theories, the theory of evolution is supported by so many observations and confirming experiments that scientists are confident that the basic components of the theory will not be overturned by new evidence.[3]

Are these comparisons logical? For instance, cell theory is based on experiments, whereas Darwin's theory of evolution is based on analogy and just-so stories. In biological studies, the theory of evolution addresses the diversity of living things, whereas cell theory addresses the uniformity of living things. In living systems, cells are the basic units of structure and function, whereas evolution is a secondary process that warrants only narrative explanations that may be right or wrong. Empirical data, delivered in 2016 at the ground-breaking British Royal Society's "New Trends in Biological Evolution" summit, revealed that the current theory of evolution is inconsistent with the observations.[4] Although a theory can be right or wrong, empirical evidence cannot be wrong. The present theory of evolution advocates competition and gradualism, whereas the empirical evidence reveals collaboration and rapidity, an outcome that appears to align with the creation model in Genesis.

These empirical results show that scientists do not understand how evolution as a secondary process propagates just as they do not understand how creation as a primary process functions. Because no human being witnessed events at the beginning of time, it is by faith that people believe in either evolutionism or creationism. No one has ever seen the force of gravity to prove that it exists, but its effect in the natural world is real and observable, whether we accept it or not. In addition, no one has ever seen the human mind or God, but their manifestations or intervention in the natural world are real and noticeable, whether we accept them or not. Hence, it is by faith that scientists trust in *chance* as the agency and discount God in their philosophical conclusion. The theory of evolution, unlike cell theory, is not a foundational scientific truth. Scientists and clerics must choose between reality and fiction.

Second, it is wrong for the church to punish any person who disagrees with its interpretation of the scriptures. It is therefore understandable that Christian clergy want to avoid the mistakes of religion in the past; but the only prudent way to do so is to stay neutral in matters of science. Their faith in evolutionists, who constitute the majority of the scientific establishment today, may be justified if no previous scientific theory has been proven wrong, but history shows that not only once but many times scientific theories such as abiogenesis or spontaneous generation, which once was accepted by a majority in science, turn out to be wrong.[5-10] No hard scientific evidence has surfaced to support the molecule-to-human being brand of evolution. Church leaders, therefore, cannot refer to creationism as a faith-based theory and to evolutionism as a fact-based theory. Without creation as a primary event, there would be no evolution as a secondary process.

Three major consequences follow from clergy's advocating the teaching only of evolutionism in public schools. First, their stance suggests that evolutionism is a state-approved precept but creationism is not. Second, clerics' endorsement of evolutionism as a foundational scientific truth opposes God's claim to have created the world in six days. This means by implication that God is fallible and hence neither omniscient nor omnipotent. Third, clerics are suggesting that evolutionism can be proclaimed/celebrated anywhere (including church premises) but God can be confined to religious premises and, hence, not omnipresent.

To create the universe in six days requires divine power, wisdom, and understanding (Jeremiah 10:12). If clerics truly believe that evolutionism and creationism are compatible as different forms of truth, that should be grounds for their comparison in science classes. When scientists portray evolutionism as truth and creationism as myth, students may question their honesty; however, when theologians insist that evolutionism is scientific truth but creationism another kind of truth based on faith, they tend to accept evolutionism as valid and brush aside creationism as worthless. No wonder, then, that the population of young Christians is dwindling while that of young atheists and agnostics is swelling.[11]

The United Methodist Church's endorsement of the Clergy Letter Project raises two other concerns. First, besides alienating creationists in their denomination, the decision of Methodist clergy risks incurring God's disapproval, who intervened in world history to ensure that belief in creationism is accepted as a moral commandment. Second, in their understandable desire to retain young people in their local congregations, the Methodist Church's official stand on evolutionism will prompt them eventually to realize the logical inconsistency and sooner or later abandon the denominational fold.

In the present Christian generation, there may not appear to be a choice between God's truth and Darwinism, but in the future, unless there is a change in the trend of events, God will be viewed by many as a delusional anachronism and Darwin as the hero of truth. Numerous clerics today are devoting one Sunday each year to a celebration of Darwin's birthday. In the years to come, many Sundays may be devoted to teaching Darwin's atheistic tenets. Because the study of science is compulsory in all public schools, it is not inconceivable that a future generation of scientists will all be essentially evolutionists with a commanding population of new atheists. God's Sabbath law on creation in the Decalogue was instituted to halt this diabolic agenda.

Pure science is neutral with regards to beliefs, and hence, has no worldview on origins but pseudoscience does. When, based on the same scientific evidence, majority of scientists and clerics advocate evolutionism and denounce creationism, they unanimously proclaim evolutionism as the state's religion. The issue is no longer a separation of church and state but as we shall see in the next chapter it becomes a matter of separation of worldviews from state policies.

7

Updating "Separation of Church and State" to "Separation of Worldviews and State Policies"

If the classroom is, indeed, as the US Supreme Court has said, "the marketplace of ideas," why not teach multiple theories regarding the origins of mankind, including Intelligent Design? Parents and their children ought to have the right to question current theories and be able to consider alternative explanations, especially when a theory is regarded as fact and has yet to be conclusively proven.... Evolution(ism) has become like a state ideology, and instead of people worrying about the separation of church and state, it has turned to an effort to become a separation of church *from* state policies.[1]

—Jack Wellman

The conviction that nothing happens supernaturally is a tenet of faith, not a fact that can be verified by any scientific means. Indeed, an *a priori* rejection of everything supernatural involves a giant, irrational leap of faith. So the presuppositions of atheistic naturalism are actually no more "scientific" than the beliefs of biblical Christianity.[2]

—John MacArthur

Prominent scholars have informed the world of the true nature and consequences of evolutionism. British zoologist Leonard Harrison Matthews (elected FRS (Fellow of the Royal Society) in 1954) said that belief in evolution parallels belief in special creation. Evolutionary biologist Clinton Richard Dawkins (elected FRS in 2001) maintained that belief in evolutionism leads to atheism. Philosopher of science Michael Ruse (elected FRSC in 1986, as a Fellow of both the Royal Society of Canada and the American Association for the Advancement of Science) declared in the *National Post* (13 May 2000, B1) that "Evolution(ism) is promulgated as an ideology, a secular religion—a full-fledged alternative to Christianity, with meaning and morality." In 2013, the world witnessed the establishment of the Atheist Mega-Church Sunday Assembly that parallels the Theist Church Sunday Assembly. These statements and events together with the pro-evolution Clergy Letter Project indicate that the boundary between church and state has evolved into a boundary between worldviews and state policies. The media and state should now be focusing on the exterminations of both creationism and evolutionism in science classrooms if the goal is not to use the advances in science to indoctrinate children and vulnerable adults.

—Michael Ebifegha

The primary goal of the pro-evolution Clergy Letter Project is to eliminate the teaching of creationism or intelligent design from public schools' science curricula on the basis of the separation of church and state, which is defined as "the distance in the relationship between organized religion and the nation state,"[3] or as "a wall of separation between Church and State."[4] The use of these phrases dates back to 1802 when Christian churches dominated America.[5] The US Supreme Court in the *Everson vs. Board of Education* case (1947) unanimously affirmed the separation of church and state by emphasizing that "the wall must be kept high and impregnable."[6] Most of the Supreme Court's Church-State decisions handed down since this case have been based on the Everson standard"[7]

Since 1947, of course, America's cultural landscape has changed radically. Most recently, the country has seen the emergence of atheistic Sunday assemblies.[8,9] The question now is thus: What exactly do we mean by "church"? Theistic and atheistic variants have different styles of worship and appeal to different worldviews. While the former broadly encompasses the Jewish, Islamic, and Buddhist faiths, it does not accommodate atheistic assemblies. The term "church" in "separation of church and state" is therefore no longer appropriate and should be replaced by "worldviews." What indoctrinates people in a sacred or secular religion is essentially its worldview. Accordingly, I would propose that "separation of church and state" be updated to "separation of worldviews and state policies." Below are some facts in support of the proposal.

First, the pro-evolution clergy are wrong in advocating for a separation of church and state. Here are the facts. Both creation and evolution are scientific terms, but supernatural creation has religious implications. The public misconception is that creation is religion, and evolution is science. Evolutionism as a worldview on origin is just as indoctrinating as creationism. Creationists in the scientific establishment are just as accomplished as evolutionists. Evolutionism is not a state religion. The debate is not between the church and state but between creationists' and evolutionists' worldviews, and the debate is about a subject (origin) that is outside science's purview. These camps of scientists are looking at the same scientific evidence and offering different philosophical interpretations. Why? Because "origin" does not have the prerequisites to qualify as a scientific topic. No one can test it; the events cannot be repeated, and no one can reproduce what exactly happened. Accordingly, evolutionist John Horgan, author of *The End of Science*, wrote, "No matter how much they learn, biologists will never really know how matter first became animate, just as cosmologists will never know how the universe began."[10] Both categories of scientists, despite a significant difference in number, have similar scientific training and qualifications, similar academic rights, and similar allegiance to the state. Besides, not all creationists believe in God or go to church; not all evolutionists disbelieve in God or do not attend church. The logical thing to do under these circumstances is to request for the separation of worldviews and state policies. Why? Because the state is made up of creationists and evolutionists; and the church is as well made up of creationists and evolutionists. How can one separate the church from the state?

Second, if clerics believe that creationism and evolutionism are two very different, but complementary, forms of truth, what is the rationale of limiting one but promoting the other in science classes? Science students have the right to compare both truths. What is truth to clergy who is not a scientist may not be truth to a science student. If, indeed, the two are complementary truths, as clerics suggest, then there is nothing left to separate the church they oversee and the state policies to which they belong.

Third, as of May 15, 2020, the ratio of the signatures of the pro-evolution Christian Clergy Letter to that of the pro-creation Christian Clergy Letter is 15,642:85 or 184:1. This shows that the pro-evolution clergy greatly outnumber the pro-creation clergy in America. Under these circumstances, what is the rationale of separating the church from the state on this matter? If the evolution worldview is superior, let it prove itself in the science classroom, and let clergy focus on their mission as servants of God and the state.

Fourth, many states' national anthems perceive God as the Creator, which means God is perceived as an essential being in these states. In acknowledging God as the Creator, these states' citizens acknowledge God as the author of science, or else how can we account for the embodiment of our scientific resources. How then can clergy or the state limit God by forbidding any reference to divinity in science classes? Again, it will only make sense to separate worldviews and the state policies.

Fifth, if God, according to Albert Einstein, is an illimitable Spirit, one would expect God's influence in both the natural and supernatural realms. Nature not only selects but also creates things at the microlevel. A field in microcreation is not pursued because it presupposes a divine Creator. The point is this: Evolution cannot precede creation, so a truly scientific worldview would recognize creation as primary causation and evolution as a subordinate process. There would then be no creationism–evolutionism controversy and, hence, no partisan bias, within the scientific establishment. Scientists would not be tagged as either creationists or evolutionists, and science would retain its disciplinary integrity. Once science is liberated from competition with organized religion, we could replace "separation of church and state" with "separation of worldviews and state policies." "Worldview" under these circumstances will be confined to theism and atheism as purely personal beliefs.

Sixth, genuine science is neither atheistic nor theistic but absolutely neutral and so has no worldview on the origin of life. The term "belief" is foreign to pure science but dominant in pseudoscience. Substituting the phrase "separation of worldviews and state policies" for "separation of church and state" consequently will have no repercussions on science, except possibly on the egos of those who think they have exclusive rights to the discipline. The United States has an obligation to support the Supreme Court's 1992 declaration that "At the heart of liberty is the right to define one's own concept of existence, meaning, of the universe, and of the mystery of human life."[11] That being the case, the state must do what it can to prevent partisanship in public education.

Modern scientists are trained to be evolutionists, so by default, we expect a majority of leading scientists to espouse an evolution worldview. There is already a well-entrenched bias on the subject of human origins, but science should be about the enterprise of objectively seeking truth rather than championing philosophical preconceptions. Because there is at present no unanimous agreement on the subject of origins, the phrase "separation of worldviews and state policies" should be binding on the teaching of all disciplines in the public sphere. The state cannot lend itself to efforts at indoctrination, particularly in science as it is being taught today. Moreover, neither religious clergy nor court judges should align themselves with the political ambitions of evolutionism as a scientific worldview. The next chapter provides guidelines on how the clergy should respond to matters that marginalize God's supremacy.

8

Seven Words of Wisdom for Clerics/Rabbis/Imams

God Almighty is the author of true religion and science and hence cannot provide two conflicting worldviews from religion and science as foundational truths.

In advocating the teaching of evolution as the only foundational scientific truth in public institutions, clerics are implying that it is religiously correct to impose scientism on the public but morally wrong to present to the public God's narrative on creation as explained and commanded in the Decalogue.

A Darwin's Day in God's church calendar! Are clerics splitting God's glory? Whether from the pulpit or elsewhere, Jehovah/Yahweh/Allah does not share glory with the dead (Isaiah 42:8). For if evolution is the primary mechanism, God, who cannot lie, would have declared such in the Decalogue.

—Michael Ebifegha

God Almighty is not pleased when clergy are lukewarm, and blend what is sacred with what is secular (Ezekiel 22:26). Promoting two conflicting worldviews as truths about origins is certainly not God's will.

Because the actual origin of nature is still unknown, science has no worldview or evidenced-based conclusions on origins— but pseudoscience does. What clerics are defending is therefore not scientific integrity but the philosophical preference of the modern scientific establishment. Charles Darwin and Alfred Russel Wallace independently and almost simultaneously developed the science of evolution by natural selection. The radical difference between these scientists was primarily philosophical and addressed the origin of man as an intellectual and moral being. Darwin believed that evolution by natural selection was responsible for both the physical and spiritual development of humankind, while Wallace believed that natural selection was responsible only for the physical development. Wallace, in *My Life: A Record of Events and Opinions*, explained:

> On this great problem the belief and teaching of Darwin was, that man's whole nature—physical, mental, intellectual, and moral—was developed from the lower animals by means of the same laws of variation and survival; and, as a consequence of this belief, that there was no difference in *kind* between man's nature and animal nature, but only one of degree. My view, on the other hand, was, and is, that there is a difference in kind, intellectually and morally, between man and other animals; and that while his body was undoubtedly developed by the continuous modification of some ancestral animal form, some different agency, analogous to that which first produced organic *life*, and then originated *consciousness*, came into play in order to develop the higher intellectual and spiritual nature

of man. This view was first intimated in the last sentence of my paper on the "Development of Human Races under Natural Selection," in 1864, and more fully treated in the last chapter of my "Essays," in 1870.

These views caused much distress of mind to Darwin, but, as I have shown, they do not in the least affect the general doctrine of natural selection. It might be as well urged that because man has produced the pouter-pigeon, the bull-dog, and the dray-horse, none of which could have been produced by natural selection alone, therefore the agency of natural selection is weakened or disproved. Neither, I urge, is it weakened or disproved if my theory of the origin of man is the true one.[1]

Darwin used natural selection to eliminate God's involvement both within the domain of science (the material or physical realm) and outside the domain of science (the immaterial or spiritual realm). Wallace, meanwhile, used natural selection to marginalize God's role in the domain of science but maintained God's involvement outside the domain. This difference is the primary reason why a predominantly atheist scientific establishment endorsed Darwinism.

I humbly request clerics to ponder for a while before answering the question below. Here are the facts:

- Evolution as a secondary natural process is inadequate to address the origin of life or species, which is a primary event.
- Science cannot locate the *mind* of an organism let alone its transformation through evolutionary mechanisms.
- Natural selection may influence diversity in the material realm. No evidence exists to suggest that it can generate diversity in the immaterial realm.

- Lucky chance (such as, nothing evolving into everything in the universe), and miracles (such as, life originating from nonlife), are not part of the evidence in pure or empirical science but are proclaimed as facts in pseudoscience only.
- Human beings, of all creatures, are the only ones who are conscious of nakedness and concerned about moral values.
- God governs or operates by laws (moral, ceremonial, civil, natural and supernatural) and in the Decalogue has claimed credit for having created life and the universe. A claim of ownership of the universe before a live audience or a nation is a civil and not a religious or scientific affair.

Here is the Question.

Based on the above facts, whose views are closer to the truth, those of Darwin or Wallace?

In the modern world, Wallace is not granted due credit because he failed to dismiss God entirely from origin science by insisting that natural selection alone could not have done it all.

True science does not demand or require any endorsement from the clergy, but pseudoscience does and that is why it requires a Clergy Letter that endorses Darwinian paradigm of evolution as a foundational scientific truth.

With due respect to the clergy of all faiths who are called by God, I humbly offer seven words of wisdom to honour the Almighty and promote truth and peace in the world.

1. Use sound judgment and remain neutral rather than take sides in a matter where God's moral law calls for the remembrance of divine creation. God knows that scientists will not understand how it took only six days to setup the world, just as we know that the gold fish in an aquarium will not understand how it took only few days to get the system running.
2. Stay within your disciplinary limits and remember that God is no respecter of persons in any controversy. Let no child in the public

or private domain go astray on account of recommendations that are inconsistent with God's will.

3. God is the source of productive knowledge and God's laws are truths, and on the subject of human origins truth cannot contradict itself. The truth of the world's creation is expressed in the Creation Sabbath law, the historical reason for our seven-day week.

4. At all times be peace-makers. Escalating disputes is Satan's job.

5. As agents of truth weigh both sides of any controversial theory in terms of its merits and not in terms of the number of its supporters. Human knowledge is not perfect, and neither are theories.

6. Remember that the Sabbath mandate in the Decalogue is a creation ordinance and not a Jewish sabbath or custom (the Jewish sabbaths such as *Day of Atonement* and *Trumpets* are excluded in the Decalogue because they are not linked to initial creation). In humility, review the Creation Sabbath law [Exodus 20:8-11], apply it to the ongoing creationism-evolutionism controversy, and reflect on the breadth and depth of God's wisdom in designating it as a sign, covenant and moral commandment. [Satan is behind the opinion that this sign, covenant and law of God on *origins* does not apply to Christian on account of Christ's atoning death on the cross. My fifth book, *Satan's Shadow in Abrahamic Religions* explains why and how Satan deceived the church.]

7. Avoid making blanket statements to the public. Stories may have figurative meanings, but all of God's moral commandments are literal. To teach otherwise is to hold the Almighty in contempt.

Here are some closing thoughts. God operates in both the natural and supernatural realms and allows everyone the free will to be a theist, agnostic, or atheist. However, in light of God's personal claim of ownership of the universe before a live audience, many nations have accepted this fact and pledged their allegiance to God in national anthems. Unfortunately, we observe a diabolical movement under the banner of science that is challenging people's free will by imposing a preferred worldview on them. The world does not celebrate a Newton Day or an Einstein Day or a Wallace Day, but it welcomes a Darwin Day in recognizing the importance of science in the betterment of humanity. Why the inconsistency?

Newton believed in a personal God; Einstein believed in a God who is not personal but illimitable and who created the laws of science; Wallace believed in natural selection but with God's involvement at the mental level. Darwin believed in natural selection with no God involved in the process. Therefore, on Evolution Sunday or Weekend, clerics are celebrating Darwin's effort to dismiss God's involvement in any stage of human life.

The institution of the Sabbath established at the Creation as recorded in the Decalogue is intended to unite the clergy and spread truth on origins to the entire world. The evolutionists within the scientific establishment have resolved to exclude God in the natural realm and are soliciting clergy and public support to silence their colleagues who are determined to follow the scientific evidence wherever it leads. Clerics' plea that science remain science and that religion remain religion forbids them to interfere in any dispute within the scientific community or celebrate a Darwin Day on religious premises. The omnipresent God declares, "Do I not fill heaven and earth?" (Jeremiah 23:24, NKJV). No one can therefore limit the illimitable God in either the natural or supernatural realm.

God bless clerics, rabbis, and imams.

Concluding Remarks

Actually, scientists are often just as prejudiced in their theories and emotionally involved in the implications of their work as are other non- scientific members of society, and are unreliable in their predictions and interpretations.[1]

—Nobel laureate Ernst Boris Chain

I have found that the most effective allies for evolution are people of the faith community. One clergyman with a backward collar is worth two biologists at a school board meeting any day![2]

—Eugenie Scott

Because science cannot resolve questions about the origin of life or species, its scholars are coming up with different philosophical interpretations of the same scientific evidence.

Science originates from God who has endowed the earth with scientific resources. Hence, one of the creation mandates is for human beings to explore the earth (Genesis 1:28). The world has been arranged in a definite order (notice the periodic table) with some degree of variation. The scientific evidence consistently points to a Creator. If the scientific evidence were unequivocally against creationism, evolutionists would invite its presentation in every classroom. The choice between theism and atheism is a matter of personal conviction, but scientific evidence is neutral. That is why opposition to Darwin's theory within the scientific establishment is coming from both theists and atheists. If it is wise not to impose religious views on people, it is also wise not to impose unproven scientific opinions on the public. A proven scientific theory is the one that is accepted by *all* scientists, not just by a majority. Let science be science, free as the evidence it reveals and without bias in any direction. Let the clergy remember that God is the embodiment of literal truth in religion and science. The notion of two very different but complementary forms of truth is misleading and devilish.

—Michael Ebifegha

Today, attendance at mainstream Christian worship services is dwindling and sustained mostly by the elderly, while the so-called Atheist Mega-Church Sunday Assembly is springing up around the globe with strong youth attendance. This development is related to the creationism-evolutionism controversy, and the damage to theism has been owing in part to the widespread acceptance of evolutionism as a kind of secular religion. This book has addressed that controversy in light of the Clergy Letter Project.

Even though the origin of life is outside science's purview, many modern scientists are determined to postulate one based solely on a paradigm in which plan, purpose, and intelligence are forbidden but random chance and mindless processes are the architects. The objective is not the discovery of truth but instead the assertion of a philosophical preference. While evolutionists in the highly secularized scientific establishment today are comfortable with this objective, creationists are prepared to settle for nothing but the truth. Evolutionists rely on their own authority, whereas creationists rely on the authority of God's revelation, which turns out to be consistent with the scientific evidence and not in need of irrational assumptions. Reference to God does not impede scientific investigation, despite evolutionists' use of this excuse to discredit creationism before the public. Sir Isaac Newton, the greatest scientist with contributions to the fields of physics, astronomy, mathematics, and chemistry, credited his discoveries to God without compromising scientific standards. He is remembered as a physicist, astronomer, mathematician, alchemist, and notably as a theologian because he wrote more volumes in religion than in science and mathematics.

The theory of evolution cannot address the matter of origins because it is based on secondary as opposed to primary causes. At best, it merely addresses the diversity of organisms but fails to provide the correct explanation of how new species arise. When the empirical evidence reveals collaboration and rapidity, the theory advocates competition and gradualism. How, then, can it be posed as a foundational scientific truth? Because of its failure to align with empirical data, honest evolutionists are abandoning the current Neo-Darwinian theory of evolution, also called the modern synthesis.

Unable to subdue their creationist and intelligent-design opponents with concrete scientific evidence, evolutionists have resorted to a divide-and-conquer approach. Eugenie Scott, director of the National Center for Science Education, has observed that clergy today are susceptible to accepting Darwinism naively. Accordingly, evolutionists such as Michael Zimmerman have appealed to liberal religious leaders for endorsement of evolutionism as the only valid and scientific worldview. Their goal was achieved with the Clergy Letter Project and the establishment of Evolution Weekends in honor of Charles Darwin. As of May 15, 2020, the Project had collected 15,642 signatures from Christian clergy. As expected, the clergy community is split with signatures from 85

Christian clergy advocating a Creation Sunday in opposition to an Evolution Sunday. It is this book's contention that clerics from both camps have stepped beyond their disciplinary boundary in endorsing or rejecting the Darwinian theory of evolution. The consequences of their folly can be gauged by the recent emergence of the Atheist Mega-Church Sunday Assembly.

Until believers honor and observe God's Creation Sabbath day as a moral and ceremonial obligation and embrace it as God's law and covenant on creation, they will continue to be divided regarding the Genesis account, and there will be no end to the creationism-evolutionism controversy. Similarly, the modern scientific establishment will remain divided on the subject of human origins until its members include microcreation, which is the predecessor of microevolution, as part of science curricula in public education. No number of court cases or Darwin Day celebrations will resolve the creationism-evolutionism controversy.

Because evolutionism like creationism is a belief that leads to atheism, it constitutes a modern religion, such that the phrase "separation of church and state" should be revised. The term "church" limits the legal principle to Christendom. Governments and courts that prohibit indoctrination in schools should therefore adopt a more universal formulation such as "separation of worldviews and state policies." Once there is no bias concerning evolution and creation worldviews in curricula, court cases will cease, and science will regain its integrity.

Appendix

Below are the details of the various Clergy Letters cited in chapter 1 each advocating the teaching of evolutionism only in public schools.

The pro-evolution Rabbi Letter (830 signatures as of May 15, 2020) reads[8]:

The Clergy Letter—from American Rabbis: An Open Letter Concerning Religion and Science

As rabbis from various branches of Judaism, we the undersigned, urge public school boards to affirm their commitment to the teaching of the science of evolution. Fundamentalists of various traditions, who perceive the science of evolution to be in conflict with their personal religious beliefs, are seeking to influence public school boards to authorize the teaching of creationism. We see this as a breach in the separation of church and state. Those who believe in a literal interpretation of the Biblical account of creation are free to teach their perspective in their homes, religious institutions and parochial schools. To teach it in the public schools would be to assert a particular religious perspective in an environment which is supposed to be free of such indoctrination.

The Bible is the primary source of spiritual inspiration and of values for us and for many others, though not everyone, in our society. It is, however, open to interpretation, with some taking the creation account and other content literally and some preferring a figurative understanding. It is possible to be inspired by the religious teachings of the Bible while not taking a literalist approach and while accepting the validity of science including the foundational concept of evolution. It is not the role of public schools to indoctrinate students with specific religious beliefs but rather to educate them in the established principles of science and in other subjects of general knowledge.

The pro-evolution Unitarian Universalist Clergy Letter (684 signatures as of May 15, 2020) reads[9]:

The Clergy Letter—from Unitarian Universalist Clergy: An Open Letter Concerning Religion and Science

As Unitarian Universalists, we draw from many sources, including "Wisdom from the world's religions which inspires us in our ethical and spiritual life," and "Humanist teachings which counsel us to heed the guidance of reason and the results of science, and warn us against idolatries of the mind and spirit." While most Unitarian Universalists believe that many sacred scriptures convey timeless truths about humans and our relationship to the sacred, we stand in solidarity with our Christian and Jewish brothers and sisters who do not read the Bible literally, as they would a science textbook. We believe that religious truth is of a different order from scientific truth. Its purpose is not to convey scientific information but to transform hearts.

Fundamentalists of various traditions, who perceive the science of evolution to be in conflict with their personal religious beliefs, are seeking to influence public school boards to authorize the teaching of creationism. We see this as a breach in the separation of church and state. Those who believe in a literal interpretation of the Biblical account of creation are free to teach their perspective in their homes, religious institutions and parochial schools. To teach it in the public schools would be to assert a particular religious perspective in an environment which is supposed to be free of such indoctrination.

We the undersigned, Unitarian Universalist clergy, believe that the timeless truths of the Bible and other scriptures may comfortably coexist with the discoveries of modern science. We believe that the theory of evolution is a foundational scientific truth, one that has stood up to rigorous scrutiny and upon which much of human knowledge and achievement rests. To reject this truth or to treat it as "one theory among others" is to deliberately embrace scientific ignorance and transmit such ignorance to our children. We urge school board members to preserve the integrity of the science curriculum by affirming the teaching of the theory of evolution as a core component of human knowledge. We ask that science remain science and that religion remain religion, two very different, but complementary, forms of truth.

The pro-evolution Buddhist Clergy Letter (75 signatures as of May 15, 2020) reads[10]:

The Clergy Letter—from American Buddhist Clergy: An Open Letter Concerning Religion and Science

"If scientific analysis were conclusively to demonstrate certain claims in Buddhism to be false, then we must accept the findings of science and abandon those claims or adopt them as metaphor."

The Universe in a Single Atom

Tenzin Gyatso - The Dalai Lama

As the above quote indicates, the Buddhist tradition is primarily a rational religion. The earliest Buddhist teachings are intended to help all sentient beings to live a life of integrity in harmony with reality.

While the specific science of evolution is not explicitly taught in our faith, it is implicit in the core teaching of interdependent origination, which demonstrates that all things are interconnected and contingent upon one another for their form and development. Likewise, a creator deity is not relied upon for a creation story. The ancient Indian fables of the Buddha's various incarnations from animal to human are readily understood not as a literal history but as metaphor describing the evolving nature of life. In fact, the concept of Buddha itself is best understood as a symbol for humanity's evolutionary potential. For all of these reasons, we admonish public school boards to affirm their commitment to teaching the science of evolution. We understand the role of public schools is to educate students in the established principles of science and in other subjects of general knowledge.

The pro-evolution Imams Letter which is no longer in circulation reads[11]:

The Clergy Letter—from American Imams: An Open Letter Concerning Religion and Science

Literalists of various religious traditions who perceive the science of evolution to be in conflict with their personal religious beliefs are seeking to influence public school boards to authorize the teaching of creationism. We, the Imams of the mosques, see this as a breach in the separation of church and state. Those who believe in a literal interpretation of scriptural account of creation are free to teach their perspective in their homes, religious institutions and parochial schools. To teach it in the public schools would be indoctrinating a particular religious point of view in an environment that is supposed to be free of such indoctrination.

We, the undersigned Imams of the mosques, assert that the Qur'an is the primary source of spiritual inspiration and of values for us, though not for everyone, in our country. We believe that the timeless truths of the Qur'an may comfortably coexist with the discoveries of modern science. As Imams we urge public school boards to affirm their commitment to the teaching of the science of evolution. We ask that science remain science and that religion remain religion, two very different, but complementary, forms of truth.

The pro-evolution Humanist Clergy Letter (54 signatures as of May 15, 2020) reads[12]:

The Clergy Letter from American Humanist Clergy An Open Letter Concerning Religion and Science

As Humanists, we have adopted a lifestance that is guided by reason, inspired by compassion, and informed by experience. Humanism is not anti-religious. It embraces a progressive philosophy which affirms our ability and responsibility to lead ethical lives of personal fulfillment that aspire to the greater good of humanity. Humanist clergy serve a growing number of individuals who variously identify as Humanists, agnostics, non-religious, and atheists, and their allies, by providing leadership, moral guidance, rites of passage, and life celebration services in a similar fashion to the clergies of other traditions.

Fundamentalists of various religions who perceive the science of evolution to be in conflict with their sectarian beliefs are seeking to influence public education authorities to require or authorize the teaching of creationism or to deprecate the teaching of evolution. We see this as a breach of the separation of church and state. Those who believe in a literal interpretation of the biblical or other religious accounts of creation are free to teach their perspectives in their homes, religious institutions, and private religious schools. But to teach creationism in its various forms, or to compromise the teaching of evolution to placate religious sensibilities, in the public schools would be to assert a particular religious perspective in an environment which is supposed to be free of such indoctrination.

We, the undersigned Humanist clergy, stand in agreement with the global scientific community that the evidence of cosmological, geological, and biological evolution is overwhelming. This consensus is in no way particular to Humanism, and we stand in solidarity with our colleagues of the Christian, Jewish, Unitarian Universalist, Muslim, and Buddhist faiths who have also embraced evolution as a vital scientific concept essential to public science education curricula. Teaching evolution in a public science classroom is no more an endorsement of Humanism than it would be of any of these otherwise disparate religious orientations.

We believe that evolution is a foundational scientific truth, one that has stood up to rigorous scrutiny and upon which much of human knowledge and achievement rests. Omitting evolution from science teaching, or treating it dismissively as "only a theory," miscommunicates its centrality in modern biology and threatens students' understanding of the very nature of science. Along with our religious allies of other traditions, we ask that science remain science and that religion remain religion. We urge public education authorities to preserve the integrity of the science curriculum by affirming the teaching of evolutionary theory as a core component of human knowledge.

References/Notes

PREFACE

1. "The Clergy Letter – from American Christian Clergy – An Open Letter Concerning Religion and Science" (http://www.theclergyletterproject.org/Christian_Clergy/ChrClergyLtr.htm). *The Clergy Letter Project.* Indianapolis, IN: Ovation Agency, Inc. Accessed December 30, 2013.

2. Creation Wiki, The Encyclopedia of Creation Science, "Creation Letter Project," http://creationwiki.org/Creation_Letter_Project Accessed December 29, 2013. Clergy for Creation - The Creation Letter - https://kcsg.wordpress.com/clergy-for-creation/ Accessed November 10, 2019.

3. Ibid.

4. Ernst Mayr, *Systematics and the Origin of Species* (New York: Columbia University Press, 1942) p.147. See also "Evolution Quotes" http://bevets.com/equotesm.htm Accessed January 17, 2014.

5. Michael Ruse, "How Evolution Became a Religion: Creationist Correct?" *National Post*, May 13, 2000, national edition, p. B1.

6. James MacAllister, "Why Neo-Darwinism was the Biggest Mistake in the History of Science" The Royal Society's Evolution Meeting in London 2016. https://evo2.org/royal-society-macallister/

7. Ibid.

8. Jerry A. Coyne, *Why Evolution Is True* (New York: Penguin Books, 2009), p. xiii.

9. Eugenie C. Scott, "My Favorite Pseudoscience" *Reports of the National Center for Science Education 23(1),* Jan/Feb 2003 http://ncse.com/rncse/23/1/my-favorite-pseudoscience Accessed May 18, 2014.

10. H.S. Lipson, "Origins of Species," *NewScientist,* 14 May 1981, p. 452.

11. Richard P. Feynman Quotes http://www.goodreads.com/author/quotes/1429989. Richard_P_Feynman Accessed May 18, 2014.

12. H.S. Lipson, "A Physicist Looks at Evolution," *Physics Bulletin* 31 (May 1980): p.138.

13. Timothy Wallace, "A theory of Creation: A Response to the Pretense that NoCreation Theory Exists" *The True.Origin Archive Exposing the Myth of Evolution*, 2000 http://www.trueorigin.org/creatheory.asp Accessed June 1, 2014.

14. Duane Gish, "The Nature of Science and of Theories on Origins, *Acts & Facts*, 24(4), 1995. http://www.icr.org/article/nature-science-theories-origin/ Accessed June 1, 2014.

15. Fr. Bill McCarthy, MSA, The Origin of "Separation of Church and State" http://www.freerepublic.com/focus/news/987191/posts Accessed May 14, 2020.

INTRODUCTION

1. H.G. Wells, *The outline of History* (NY, Doubleday & Company, 1961), p. 776.

2. George Murphy' "Re: [asa] Jonathan Wells essay" January 29, 2007, http://www2.asa3.org/archive/asa/200701/0634.html Accessed January 12, 2014.

3. Ibid.

4. Jerry A. Coyne, *Why Evolution Is True* (New York: Penguin Books, 2009) p. 223.

5. Ibid. p. 249.

6. Ibid. xiii.

7. Douglas H. Erwin, "Darwin Still Rules, but Some Biologists Dream of a Paradigm Shift" *The New York Times*, Tuesday, June 26, 2007, p. F2. Modern Synthesis is simply the blending of natural selection and genetic variation.

8. Evolution News and Views, "Junk No More: ENCODE Project *Nature* Paper Finds "Biochemical Functions for 80% of the Genome" http://www.evolutionnews.org/2012/09/junk_no_more_en_1064001.html Accessed January 2, 2014.

9. "The Shape of Evolution: A Comparison of Real and Random Clades," *Paleobiology* 3(1) (1977), pp. 34-35.

10. Michael N. Marcus, 1992: Catholic Church apologizes to Galileo, who died in 1642. http://4thefirsttime.blogspot.ca/2007/09/1992-catholic-church-apologizes-to.html

11. Jerry Fodor and Massimo Piattelli-Palmarini, *What Darwin Got Wrong* (New York: Farrar, Straus, and Giroux, 2010), p. xx.

12. "The Clergy Letter – from American Christian Clergy – An Open Letter Concerning Religion and Science" (http://www.theclergyletterproject.org/Christian_Clergy/ChrClergyLtr.htm). *The Clergy Letter Project*. Indianapolis, IN: Ovation Agency, Inc. Accessed December 30, 2013.

13. *Why Evolution Is True* (New York: Penguin Books, 2009) p. xiii.

14. Pierre Teilhard de Chardin, *The Phenomenon of Man* (New York: Harper Perennial, 2002), pp. 218–219.

THE EVOLUTION AND CREATION CLERGY LETTER PROJECT

1. Dr. Jill Joseph, "A Different Way of Knowing-All Saints Episcopal Church" Evolution Weekend Sermon, February 9, 2014 http://www.allsaintssacramento. org/sermon-overview/different-way-knowing Accessed April 12, 2014.
2. Clergy Letter Project: http://en.wikipedia.org/wiki/Clergy_Letter_Project Accessed December 30, 2013. Michael Zimmerman, The Clergy Letter Project: https:// www.theclergyletterproject.org/ Accessed November 10, 2019.
3. Andy Coghlan, "American Muslim clerics sign up for evolution, May 2011 http:// www.newscientist.com/article/dn20522-american-muslim-clerics-sign-up-for-evoluti... Accessed December 30, 2013.
4. Michael Zimmerman, Ph.D., *The Huffington Post.* http://www.huffingtonpost. com/michael-zimmerman/
5. Jeff Nall, "Conversations with Christian and Atheist Activists: Michael Zimmerman," *American Humanist Association*: *Humanist Network News Ezine Archives.* http:// www.americanhumanist.org/hnn/archives/index. php?id=278&article=1 Accessed December 30, 2013.
6. Creation Wiki, The Encyclopedia of Creation Science, "Creation Letter Project," http://creationwiki.org/Creation_Letter_Project Accessed December 29, 2013. Clergy for Creation - https://kcsg.wordpress.com/clergy-for-creation/ Accessed November 10, 2019.
7. "The Clergy Letter – from American Christian Clergy – An Open Letter Concerning Religion and Science" (http://www.theclergyletterproject.org/ Christian_Clergy/ChrClergyLtr.htm). *The Clergy Letter Project.* Indianapolis, IN: Ovation Agency, Inc. Accessed May 15, 2020.
8. "The Clergy Letter – from American Rabbis – An Open Letter Concerning Religion and Science" (http://www.theclergyletterproject.org/Jewish_Clergy/ JewishClergyLtr. htm). *The Clergy Letter Project.* Indianapolis, IN: Ovation Agency, Inc. May 15, 2020.
9. "The Clergy Letter – from Unitarian Universalist Clergy – An Open Letter Concerning Religion and Science" (http://www.theclergyletterproject.org/ Unitarian_Universalists/UnivUnitarianClergyLtr.htm). *The Clergy Letter Project.* Indianapolis, IN: Ovation Agency, Inc. Accessed May 15, 2020.
10. "The Clergy Letter – from American Buddhist Clergy – An Open Letter Concerning Religion and Science" (http://theclergyletterproject.org/Buddhist_ Clergy/BuddhistClergyLtr.html). *The Clergy Letter Project.* Indianapolis, IN: Ovation Agency, Inc. Accessed May 15, 2020.
11. The Clergy Letter – from American Imams – An Open Letter Concerning Religion and Science" (http://theclergyletterproject.org/ImamLetter.html). *The*

Clergy Letter Project. Indianapolis, IN: Ovation Agency, Inc. Accessed December 30, 2013 through the *NewScientist* website in reference 2 above.

12. "The Clergy Letter– from American Humanist Clergy – An Open Letter Concerning Religion and Science" (https://theclergyletterproject.org/Humanists/HumanistsLetter.html Accessed May 15, 2020.

13. Creation Wiki, The Encyclopedia of Creation Science, "Creation Letter Project," http://creationwiki.org/Creation_Letter_Project Accessed May 16, 2020.

14. Ken Ham, Hugh Ross, Deborah B. Haarsma, Stephen C. Meyer, *Four Views on Creation, Evolution, and Intelligent Design,* (Grand Rapids, Michigan: Zondervan, 2017), pp.11-13.

SEVEN TRUTHS IN THE EVOLUTIONISM-CREATIONISM CONTROVERSY

1. David Shiga, "Earth may have had water from day one," *New Scientist* (November 05, 2010), http://www.newscientist.com/article/mg20827853.800-earth-may-have-had-water-from-day-one.html. Accessed November 07, 2010.

2. *The Darwinian Delusion: The Scientific Myth of Evolutionism* (AuthorHouse, Bloomington, IN: AuthorHouse, 2011, pp. 78-100.

3. *What Evolution Is* (New York: Basic Books, 2001), p. 190.

4. Special to the *New York Times*, "Obituary—Einstein Noted as an Iconoclast in Research, Politics, and Religion," *New York Times*, April 19, 1955, p. 25.

5. The Archives of the Episcopal Church, "Affirm Evolution and Science Education": Resolution Number 2006-A129, General Convention, *Journal of the General Convention of...The Episcopal Church, Columbus, 2006* (New York: General Convention, 2007), pp. 690-691. (http://www.episcopalarchives.org/cgi-bin/acts/acts_resolution-complete.pl?resolution=2006-A129) Accessed December 31, 2013.

6. The Origin-of-Life Prize, About the Gene Emergence Project,, http://www.us.net/life/rul_abou.htm; The Origin-of-Life Prize ®, Description of the Prize, http://www.us.net/life/rul_desc.htm; The Origin-of-Life Prize ®, Prize Value, http://lifeorigin.org/rul_priz.htm; The Origin-of-Life Prize ®, Submissions, http://lifeorigin.org/rul_late.htm The Origin-of-Life Prize ®, Late News: http://lifeorigin.org/rul_subm.htm. Accessed March 30, 2014.

7. The Origin-of-Life Prize, "lifeorgin-info," http://www.us.net/life/rul_late.htm Accessed January 18, 2014.

8. The Origin-of-Life Prize ®, Discussion, http://lifeorigin.org/rul_disc.htm. Accessed April 5, 2014.

9. http://www.us.net/life/rul_abou.htm.

10. "Social Responsibility and the Scientist," *Perspectives in Biology and Medicine in Modern Western Society* 14 (1971), pp. 367-368.

11. Brian Thomas, "30 Years Later, the Lessons from Mount St. Helens." http://www.icr.org/article/5465/ Accessed January 18, 2014.

12. Andy Coghlan, "With no paper trail, can science determine age? http://www.newsientist.com/article/mg21428644.300-with-no-paper-trail-can-science-deter... Accessed June 6, 2013.

13. Stephen Hawking and Leonard Mlodinow, *The Grand Design* (New York: Bantam Books, 2010), 124.

14. Wikipedia, the free encyclopedia, Week http://en.wikipedia.org/wiki/ Week The Old Farmer's Almanac, Why the Week Has Seven Days (1988) http://www. almanac.com/content/why-week-has-seven-days Accessed May 26, 2014.

15. Madeleine J. Nash, "When Life Exploded," *Time,* December 4, 1995, 66–69; Jeffrey S. Levinton, "The Big Bang of Animal Evolution," *Scientific American* 267, no.5 (November 1992): 84–91; AllAboutScience.org, "The Cambrian Explosion—Biology's Big Bang?" (http://www.allaboutscience.org/the-cambrian- explosion. htm), Accessed January 1, 2014.

16. Wikipedia, the free encyclopedia, Uniformitarianism, http://en.wikipedia.org/ wiki/Uniformitarianism Accessed March 11, 2014.

17. L. Harrison Matthews, introduction to *The Origin of Species*, by Charles Darwin (London: reprinted by J. M. Dent & Sons Ltd., 1971), pp. x–xi.

18. On Methods of Evolutionary Biology and Anthropology," *American Scientist* 45 (1957), p. 388.

19. H.S. Lipson, "A Physicist Looks at Evolution," *Physics Bulletin* 31 (May 1980), p. 138.

20. Chengjiang Maotianshan Shales Fossils". Fossil Mall: Science Section (http://www.fossilmall.com/Science/Sites/Chenjiang/Chengjiang.htm) Accessed February 15, 2014.

21. Wikipedia, the free encyclopedia, Chengjiang County, http://en.wikipedia.org/ wiki/Chengjiang_County Accessed February 15, 2014.

22. Sterelny, Kim *Dawkins vs. Gould: Survival of the Fittest.* (Cambridge, U.K.: Icon Books, 2007) p. 116.

23. Jerry A. Coyne, *Why Evolution Is True* (New York: Penguin Books, 2009), p. 223.

24. E. J. H., Corner, "Evolution," *Contemporary Botanical Thought*, ed. Anna M. MacLeod and L. S. Cobley (Edinburgh: Oliver and Boyd, 1961), p. 97.

25. Mark Ridley, "Who Doubts Evolution?" *NewScientist* 90 (1981), p. 831.

26. The Origin-of-Life Prize ®, Discussion, http://lifeorigin.org/rul_disc.htm. Accessed April 5, 2014.

27. *Why Evolution Is True,* first and second pages.

28. Ibid. 223-224--Bad Design is found on pp. 81-85.

29. The ENCODE Project Consortium, "An integrated encyclopedia of DNA elements in the human genome, *Nature*, Vol. 489, September 6, 2012, pp. 57-74.

30. Evolution News and Views, "Junk No More: ENCODE Project *Nature* Paper Finds "Biochemical Functions for 80% of the Genome" http://www.evolutionnews. org/2012/09/junk_no_more_en_1064001.html Accessed January 2, 2014.

31. David Klinghoffer, In Debate, Britain's Chief Rabbi Tweaks Richard Dawkins with the Myth of "Junk DNA," "Evolution News And Views," http://www. evolutionnews.org/2012/09/in_debate_brita_1064521.html Accessed March 10, 2014.

32. Richard Dawkins, *The Greatest Show On Earth: The Evidence For Evolution* (New York: Free Press, 2009), pp. 332-333.

33. Ibid. first page.

34. Scott C. Todd, "A View from Kansas on That Evolution Debate," *Nature*, 30 September 1999, p. 423.

35. James Reddie, "Scientia Scientiarum," *Journal of the Transactions of the Victoria Institute* (1865), http://www.creationism.org/victoria/VictoriaInst1866_pg005. htm Accessed January 2, 2014.

WHY EVOLUTION(ISM) IS NOT A FOUNDATIONAL SCIENTIFIC TRUTH

1. James MacAllister, "Why Neo-Darwinism was the Biggest Mistake in the History of Science" The Royal Society's Evolution Meeting in London 2016. https://evo2. org/royal-society-macallister/James MacAllister curator of the Lynn Margulis archive at the University of Massachussetts-Amherst, wrote a thorough review of the London 2016 evolution conference at the Royal Society.

2. Agard, E T., in *In six days — Why 50 Scientists choose to believe in Creation*, John FX Ashton (ed.), New Holland Publishers, Sydney, 1999, pp. 196-197.

3. Jerry Fodor and Massimo Piattelli-Palmarini, *What Darwin Got Wrong* (New York: Farrar, Straus, and Giroux, 2010), p. 163.

4. Darwin C., *The Origin of Species* (London: Penguin Books, 1985), 455.

5. Darwin C., *The Origin of Species* (New York: Penguin Edition, 1968), 291-292.

6. John Graham, Response to"Why Neo-Darwinism was the Biggest Mistake in the History of Science." The Royal Society's Evolution Meeting in London 2016.

7. John Hands, Comments in "Why Neo-Darwinism was the Biggest Mistake in the History of Science" The Royal Society's Evolution Meeting in London 2016.

8. Sterelny, Kim *Dawkins vs. Gould: Survival of the Fittest.* (Cambridge, U.K.: Icon Books, 2007) p. 116.

9. Arthur Miller, Royal Society's "New Trends in Biological Evolution" — A Bloodless Revolution, November 30, 2016. https://evo2.org/royal-society-evolution/ Accessed April 27, 2020.

10. Jerry Fodor and Massimo Piattelli-Palmarini, *What Darwin Got Wrong* (New York: Farrar, Straus, and Giroux, 2010), p. xx.

11. John Horgan, "Was Darwin Wrong? A journalist recounts the epic story of modern challenges to evolutionary dogma." June 9, 2019.

12. Ibid.

13. Luskin C. "Design vs. Descent: A Contest of Predictions" IDEA Center, 2019. http://www.ideacenter.org/contentmgr/showdetails.php/id/846 Accessed December 2, 2019.

14. Lewontin, R., "Billions and Billions of Demons", The New York Review, Jan 9, 1997, p. 31.

15. Todd, Scott, C., "A view from Kansas on That Evolution Debate", *Nature*, Vol. 401, No. 6752, 30 September 1999, p. 423.

16. Royal Society Announcement: $10 Million Prize for Life's Origin. https://evo2.org/royalsociety/ Accessed November 24, 2019.

17. The Origin-of-Life Prize. Discussion. http://www.us.net/life/rul_defi.htm.

18. S. Powell, "The universe may be a billion years younger than we thought. Scientists are scrambling to figure out why." May 18, 2019. https://www.nbcnews.com/mach/science/universe-may-be-billion-years-younger-we-thought-scientists-are-ncna1005541 Accessed November 24, 2019.

19. Seth Borenstein, "Study finds the universe might be 2 billion years younger" September 12, 2019. https://apnews.com/4f9cd2357365418e81104a33994c0d32 A

20. Nancy E. Spaulding, Samuel N. Namowitz, *Earth Science* (Illinois: McDougal Littell Inc., 2003), 650.

21. Alfredo Carpineti, "Where Did Earth's Water Come From?" IFLScience, 12 November 2015. http://www.iflscience.com/physics/origin-earths-water-discovered-0/ Accessed September 16, 2017.

22. David Shiga, "Earth may have had water from day one," *New Scientist* (November 05, 2010), http://www.newscientist.com/article/mg20827853.800-earth-may-have-had-water-from-day-one.html. Accessed November 07, 2010.

23. Metro News, "Early Earth 'was covered in water' *Metrowebukmetro* December 31, 2008 http://metro.co.uk/2008/12/31/early-earth-was-covered-in-water-274995/ Accessed October 12, 2017.

24. Dr. Andrew A. Snelling, Dr. Joe Francis, and Tom Hennigan, "Lasting Lessons from Mount St. Helens" April 1, 2015. https://answersingenesis.org/geology/mount-st-helens/lasting-lessons-mount-st-helens/ Accessed September 16, 2017.

25. Brian J. Alters, Sandra M. Alters, *Defending Evolution, in the Classroom* (London: Jones and Bartlett, 2001), 119.

26. Blueink Review, *Creation or Evolution?: Origin of Species in Light of Science's Limitations and Historical Records,* by Michael Ebifegha, Review Date November 2011. https://www.blueinkreview.com/book-reviews/creation-or-evolution-origin-of-species-in-light-of-sciences-limitations-and-historical-records/

27. Planet Dinosaur, "Dinosaur Anatomy – Dinosaur Tails" http://planetdi.startlogic.com/dinosaur_anatomy/tails.htm

28. The Virtual Fossil Museum, "Fossils Across Geological Time and Evolution, 1999. http://www.fossilmuseum.net/GeologicalHistory.htm:

29. Wikipedia, "Geologic time scale – Epic of Evolution http://epicofevolution. com/GeoTimeWiki.html, Accessed September 25, 2017.

30. Dolores, R. Piperno, and Hans-Dieter Sues, "Dinosaurs Dined on Grass", *Science,* Vol. 310, (November 18, 2005): 1126.

31. Mary H. Schweltzer, Jennifer L. Wittmeyer, John R. Horner, Jan K. Toporski, "Soft-Tissue Vessels and Cellular Preservation in Tyrannosaurus rex", *Science,* Vol. 307, (March 25, 2005): 1952-1955.

32. Mary H. Schweltzer, "Blood from Stone," *Scientific American* 303, No. 6 (December 2010): 62-8.

33. Max Jammer, *Einstein and Religion: Physics and Theology* (Princeton, NJ: Princeton University Press, 1999), 96-97.

34. Ibid. 97.

35. Perry Marshall, "83-Year-Old Heart Scientist Rocks the Foundations of Evolution," This article was sent to my email on October 5, 2020. Perry Marshall is the author of *Evolution 2.0: Breaking the Deadlock Between Darwin and Design*. Contact: Perry Marshall - EV2.0 (info@evo2.org).

FOLLIES IN THE EVOLUTION AND CREATION CLERGY LETTERS

1. Joel Watts, "Unsettled Christianity" What Does The Clergy Letter Project Really Say? March 5, 2011. http://unsettledchristianity.com/2011/03/what-does-the-clergy-letter-project-really-say/ Accessed February 17, 2014.

2. "Evolution's Final Frontiers: 16 Challenges for a Modern-Day Darwin," *New Scientist,* 31 January 6 – February 6, 209. pp. 41-43.

3. "The Evolutionary Vision," *Issues in Evolution,* vol. 3, ed. Sol Tax and Charles Callender (Chicago: University of Chicago Press, 1960), pp. 252-253.

4. *Tornado in a Junkyard* (Arlington, MA: Refuge Books, 1999), p. 233.

5. *Science, Evolution, and Creationism* (Washington, D.C.: National Academies Press, 2008), p. 12.

6. *Creation: Facts of Life* (Green Forest, AR: Master Books, 2006), p. 11.

7. *The Blind Watchmaker* (London: Penguin, 2006), p. xv.

8. *From Evolution to Creation: A Personal Testimony* (Florence, KY: Answers in Genesis, 2000), p. 10.

9. *From Evolution to Creation.* pp. 12-13.

10. Ibid. p. 7. Dr. Gary Parker is an elected member of the national university scholastic honorary society Phi Beta Kappa, recipient of two nationally competitive fellowship awards, and principal author of five programmed biology textbooks published by John Wiley and Sons.

11. *Origins: Linking Science and Scripture* (Hagerstown, MD: Herald Publishing, 1998), p. 182.

12. Jerry Bergman: *Slaughter of the Dissidents: The Shocking Truth about Killing the Careers of Darwin Doubters* (Leafcutter Press, 2008, 477 pages). http://www.godofcreation.com/essays/display.asp?ind=92 Accessed January 12, 2014.

13. *Wikipedia, the free encyclopedia* "Clergy Letter Project" http://en.wikipedia.org/wiki/Clergy_Letter_Project Accessed January 25, 2014.

14. The Clergy Letter Project, "2014 Evolution Weekend" http://www.theclergyletterproject.org/rel_evolution_weekend_2014.html Accessed January 12, 2014.

15. PZ Myers, "Evolution Sunday?", February 11, 2007. http://scienceblogs.com/pharyngula/2007/02/evolution_sunday.php Accessed January 23.

16. Michael Zimmerman, *The Huffington Post*: http://huffingtonpost.com/michael-zimmerman/ Accessed January 26, 2014.

17. Pierre-Paul Grassé, *Evolution of Living Organisms,* (New York: Academic Press, 1977), pp. 106-107.

18. Max Jammer, *Einstein and Religion: Physics and Theology* (Princeton, NJ: Princeton University Press, 1999), 48.

19. Ibid. 122–23.

20. "Evolution, Creation, and Biology Teaching," *Evolution Versus Creationism: The Public Education Controversy,* ed. J. P. Zetterberg (Phoenix: Oryx Press, 1983), p. 91.

21. Teaching of Origins; Letter to the Secretary of State for Education regarding the teaching of origins http://www.biblicalcreation.co.uk/educational_issues/bcs116. html Accessed January 27, 2014. According to The Biblical Creation Society, the letter was reported in the *Times Educational Supplement* of April 26, 2002, page 14, by Clare Dean. The title of her article was *"Let's Teach Science Pupils How To Think"* The signatories include:
    * Andy McIntosh DSc, FIMA, CMath, FInstE, CEng Professor of Thermodynamics and Combustion Theory, University of Leeds.
    * Edgar Andrews BSc, PhD, DSc, FInstP, FIM, CEng, CPhys. Emeritus Professor of Materials Science, University of London.
    * D B Gower BSc, PhD, DSc, CChem, FRSC, CBiol, FIBiol Emeritus Professor of Steroid Biochemistry, University of London.

22. Jerry Fodor and Massimo Piattelli-Palmarini, *What Darwin Got Wrong* (New York: Farrar, Straus, and Giroux, 2010), p. xiii.

23. *There Is a God: How the World's Most Notorious Atheist Changed His Mind* (New York: HarperCollins, 2007), p. 88-89.

24. Ibid. p. *93.* Ibid. p. 93.

25. Max Jammer, *Einstein and Religion: Physics and Theology,* (Princeton NJ: Princeton University Press, 1999), pp. 93-94.

26. Richard Lewontin, "Billions and Billions of Demons", *New York Review of Books;* January 9, 1997, p. 31.

27. Francis Crick, *What Mad Pursuit: A Personal View of Scientific Discovery,* (New York: Basic Books, 1990), p. 138.

28. L. C., Birch and, P. R., Ehrlich, "Evolutionary History and Population Biology", *Nature,* 214, (1967), p. 352.

29. Jerry A. Coyne, *Why Evolution Is True* (New York: Penguin Books, 2009) pp 66-68.

30. The Clergy Letter – from American Christian Clergy – An Open Letter Concerning Religion and Science" (http://www.theclergyletterproject.org/ Christian_Clergy/ChrClergyLtr.htm).

31. William B. Provine, "Evolution Quotes" http://bevets.com/equotesp5.htm Accessed January 12, 2013. "No free Will" in *Catching up with the Vision,* Margaret W. Rossiter (Ed.), Chicago University Press, p. S123, 1999. CreationRevolution, "William Provine on Evolution and Atheism" http://creationrevolution. com/2011/04/william-provine-on-evolution-and-atheism/ Accessed January 12, 2014.

32. Michael Dowd, *Thank God for Evolution* (New York: Plume, 2009).

33. Michael Dowd, "Celebrating Evolution. Michael Dowd to Christian Church: New Atheists are God's Prophets" Sunday/ August 1, 2010. http://michaeldowd.org/ media/files/82890dc8b7595fd71432ebf3f4d4f391-74.php Accessed June 9, 2013.

34. The Clergy Letter – from American Christian Clergy.

35. Ken Ham, "Churches in praise of… Darwin! February 8, 2007. http://www. answersingenesis.org/articles/2007/02/08/churches-praise-darwin&vPrint=1 Accessed May 12, 2013.

36. John Allemang, "The infinite wisdom of Richard Dawkins" *The Globe and Mail,* Toronto, June 23, 2007. http://old.richarddawkins.net/articles/1324-the-infinite-wisdom-of-richard-dawkins Accessed June 9. 2013.

37. Ariel A. Roth, *Origins: Linking Science and Scripture* (Hagerstown, MD: Herald Publishing, 1998), p. 17.

38. Richard Dawkins, *The Selfish Gene* (Oxford: Oxford University Press, 2006), pp. 19-20.

39. Denis Noble, *The Music of Life: Biology Beyond The Genome* (Oxford: Oxford University Press, 2006), p. 12.

40. *There Is a God: How the World's Most Notorious Atheist Changed His Mind,* p. 89.

41. Ibid, pp. 90-91.

42. Wikipedia: Large Hadron Collider http://en.wikipedia.org/wiki/ Large_Hadron_Collider Accessed December 3, 2013.

43. A. E. Wilder-Smith, *The Creation of Life* (Costa Mesa, TWFT Publishers, 1988), p. 228. Wilder-Smith (D.Sc., Ph.D., Dr. es. Sc., F.R.I. C), was a scholar of natural science, Oxford and Reading Universities, England.

44. Jonathan Wells, The Politically Incorrect Guide to Darwinism and Intelligent Design (Washington, DC: Regnery Publishing, 2006), p. 82.

45. Pierre Teilhard de Chardin, *The Phenomenon of Man* (New York: Harper Perennial, 2002), pp. 218–219.

46. Hubert Reeves, Joel De Rosnay, and Yves Coppens, *Origins: Cosmos, Earth, and Mankind* (New York: Arcade Publishing, 1998), p. 78.

47. The Origin of Life," *Scientific American* 191.2 (1954), p. 46.

48. "Life and Mind in the Universe," *Cosmos, Bios, Theos*, ed. Henry Margenau and Roy Abraham Varghese (La Salle, IL: Open Court, 1992), pp. 218-19.

49. Rev. Kyle. Ann Lovett Church of the Crossroads, Honolulu, Hawai "Sermon Notes for February 9, 2014 – Out of the Muck, . http://churchofthecrossroadshawaii.org/wp-content/uploads/sites/29/2014/03/2014-February-9-Sermon-Out-of-the-Muck.pdf

50. Christians Answers Network, "Should the book of Genesis be taken literally? https://www.christiananswers.net/q-aig/aig-c020.html Accessed May 3, 2014.

51. https://www.christiananswers.net/q-aig/aig-c024.html Accessed May 3, 2014.

52. *Creation or Evolution? Origin* of Species in *Light of Science's Limitations and Historical Records* "Kirkus Review" http://www.kirkusreviews.com/book-reviews/michael-ebifegha/creation-or-evolution

53. John Rennie, "15 Answers to Creationist Nonsense," *Scientific American* 287, (1): July 2002, p. 80.

54. Jonathan Sarfati, 15 Creationist Answers to Unscientific American John Rennie, *Crossroads of Religion* June 25, 2007. http://religiouscrossroads.tribe.net/ thread/2e951110-6032-402a-8d05-267c4835f14c Accessed May 19, 2014.

55. Do-While Jones "No Nonsense" *Science Against Evolution* http://www.ridgenet.net/~do_while/sage/v6i10f.htm Accessed May 19, 2014.

ANALYSIS OF A PRO-EVOLUTION SERMON

1. "The Clergy Letter – from American Christian Clergy – An Open Letter Concerning Religion and Science" (http://www.theclergyletterproject.org/Christian_Clergy/ ChrClergyLtr.htm). *The Clergy Letter Project*. Indianapolis, IN: Ovation Agency, Inc. Accessed May 15, 2020.

2. CBS News, "Athiest "mega-churches" take root across U.S., world" November 11, 2013. https://www.cbsnews.com/news/atheistmega-churches-take-root-across-us-world/ Accessed May 18, 2020.

3. Dr. Jill Joseph, "A Different Way of Knowing-All Saints Episcopal Church" Evolution Weekend Sermon, February 9, 2014 http://www.allsaintssacramento.org/sermon-overview/different-way-knowing Accessed April 12, 2014.

4. *Webster's Universal Dictionary & Thesaurus,* (Scotland: Geddes & Grosset, 2003).

5. Pierre-P Grassé, *Evolution of Living Organisms: Evidence for a New Theory of Transformation* New York: Academic Press, 1977), p. 8.

6. Dr. Ralph Walker, Emeritus Fellow and Lecturer in Philosophy, in Magdalen College Chapel, Oxford; A sermon for 'Evolution Sunday' (the Sunday nearest Charles Darwin's birthday) preached on 9 February 2014. http://www.magd.ox.ac.uk/wordpress/wp-content/uploads/2013/07/9-February-2014.pdf Accessed April 19, 2014.

7. Tim Chaffey, *Answers in Genesis*-U.S., Feedback: "Do Genesis 1 and 2 Contradict Each Other? http://www.answersingenesis.org/articles/2010/09/03/feedback-genesis-1-and-2 Accessed April 19, 2014.

8. Dr. Ralph Walker, Evolution Sunday, 9 February 2014. http://www.magd.ox.ac.uk/wordpress/wp-content/uploads/2013/07/9-February-2014.pdf Accessed May 26, 2014.

9. Watch Tower, Bible and Tract Society of New York, (Brooklyn, New York, 1998) p. 93.

CONSEQUENCES OF THE PRO-EVOLUTION CLERGY LETTER PROJECT

1. Francis S. Collins, *The Language Of God* (New York: Free Press, 2006), 176.

2. Albert C. "Al" Kuelling Special Contributor in *United Methodist Portal* "Methodism supports teaching of evolution" *The United Church Portal*, July 31, 2008 http://www.umportal.org/article.asp?id=3869 Accessed August 5, 2013.

3. National Academy of Sciences and Institute of Medicine, *Science, Evolution, and Creationism*, (Washington, DC: The National Academies Press, 2008), p.11.

4. James MacAllister, "Why Neo-Darwinism was the Biggest Mistake in the History of Science" The Royal Society's Evolution Meeting in London 2016. https://evo2.org/royal-society-macallister/

5. Top 10: Disproven Theories http://ca.askmen.com/top_10/entertainment/top-10-disproven-theories_10.html Accessed April 21, 2014. These include: Spontaneous generation, Geocentric universe, Phlogiston, Catas-trophism.

6. Eric Berger, *Houston Chronicle*, "The top 10 most spectacularly wrong widely held scientific theories November 24, 2010, http://blog.chron.com/sciguy/2010/11/the-top-10-most-spectacularly-wrong-widely-held-scientific-theories/ Accessed April 21, 2014.

7. Evan Andrews, *TopTenz* "Top 10 Most Famous Scientific Theories (That Turned out to be Wrong) March 12, 2010. http://www.toptenz.net/top-10-most-famous-scientific-theories-that-turned-out-to-be-wrong.php Accessed April 21, 2014.

8. Jamie Frater, "Ten Debunked Scientific Beliefs Of The Past- Listverse, January 19, 2009. http://listverse.com/2009/01/19/10-debunked-scientific-beliefs-of-the-past/ Accessed August 6, 2013.

9. *Creation or Evolution: Origin of Species in Light of Science's Limitations and Historical Records* (Bloomington, IN: iUniverse, 2011), pp. 92-99.

10. Michael Denton, *Evolution:* A Theory in Crisis (Bethesda, MD: Adler & Adler, 1986), pp. 384-351.

11. Gillian Flaccus "Atheist 'Megachurches' Crop Up Around The World" 11/10/13. http://www.huffingtonpost.com/2013/11/10/atheist-mega-church_n_4252360. html Accessed May 27, 2014.

UPGRADING "SEPARATION OF CHURCH AND STATE" TO "SEPARATION OF WORLDVIEWS AND STATE POLICIES"

1. Jack Wellman, Yahoo Contributor Network, Separation of Church and State and Public Schools Science Curriculum: Intelligent Design Vs Evolution: Separation of Church from State, Feb 15, 2010. http://voices.yahoo.com/separation-church-state-public-schools-5479947.html?cat=9 Accessed March 4, 2014.

2. John MacArthur, *The Battle for the Beginning,* (Nashville: W Publishing Group, 2001), pp. 50-51.

3. Wikipedia, the free encyclopedia; "Separation of Church and state", http://en.wikipedia.org/wiki/Separation_of_church_and_state Accessed April 21, 2014.

4. Fr. Bill McCarthy, MSA, The Origin of "Separation of Church and State" http://www.freerepublic.com/focus/news/987191/posts Accessed May 14, 2020.

5. David Brancaccio, "NOW" Politics & Economy, God and Government. Separating Ch… http://www.pbs.org/now/politics/churchandstate.html Accessed April 21, 2014.

6. Ibid.

7. Ibid.

8. Gillian Flaccus, *Yahoo News,* "Atheist 'mega-churches" take roots across US, world, November 10, 2013. http://news.yahoo.com/atheist-mega-churches-root-across-us-world-214619648. html Accessed April 21, 2014.

9. *The Guardian,* "Atheist Sunday Assembly branches out in first wave of expansion" http://www.theguardian.com/world/2013/sep/14/atheist-sunday-assembly-branches-out Accessed April 23, 2014.

10. John Horgan, "Was Darwin Wrong? A journalist recounts the epic story of modern challenges to evolutionary dogma." June 9, 2019.

11. Fr. Bill McCarthy, MSA, The Origin of "Separation of Church and State" http://www.freerepublic.com/focus/news/987191/posts Accessed May 14, 2020.

SEVEN WORDS OF WISDOM FOR CLLERIES/RABBIS/IMAMS

1. Alfred Russell Wallace, *My Life: A Record of Events and Opinions,* (London: Chapman & Hall, 1905), Volume II, p. 17.

CONCLUDING REMARKS

1. Ernst Boris Chain, "Social Responsibility and the Scientist," *Perspectives in Biology and Medicine* 14 (1971), p. 353.
2. Thomas J. Oord and Eric Stark, A Conversation with Eugenie Scott, Science and Theology News, February 2004. http://web.archive.org/web/20050309214648/ http://www.stnews.org/archives/2002/Apr_features.html#2 Accessed June 6, 2014.

APPENDIX

Details of the references 8-12 cited in Chapter 1 are provided here.

About the Author

Michael Ebifegha, an ecumenical Catholic/Christian, is a certified science, mathematics, and religion instructor. Ebifegha earned a certificate in religious studies from the Toronto Catholic School Board, a Bachelor of Education in science and mathematics, and a Ph.D. in physics from the University of Toronto. He is the author of *Farewell to Darwinian Evolution: God's Creation Patent and Seal; The Darwinian Delusion: The Scientific Myth of Evolutionism; Creation or Evolution? Origin of Species in Light of Science Limitations and Historical Records; and Satan's Shadow in Abrahamic Religions: Clerics' defiance of God's Creation Sabbath Day mandate in celebrating Charles Darwin's Evolution Day in their places of worship.*

www.ingramcontent.com/pod-product-compliance
Lightning Source LLC
Chambersburg PA
CBHW071353120626
46546CB00002B/672